2 Progress in

Clinical Biochemistry and Medicine

Oncogenes and Human Cancer
Blood Groups in Cancer
Copper and Inflammation
Human Insulin

With Contributions by
T. L. J. Boehm, U. Deuschle
W. J. Kuhns, R. Obermeier, F. J. Primus
U. Weser, M. Zoltobrocki

With 25 Figures

Springer-Verlag
Berlin Heidelberg New York Tokyo

As a rule, contributions to this series are specially commissioned. The editors and publishers will, however, always be pleased to receive suggestions and supplementary information. Papers are accepted for "Progress in Clinical Biochemistry and Medicine" in English.

ISBN-13: 978-3-642-70572-4 e-ISBN-13: 978-3-642-70570-0
DOI: 10.1007/978-3-642-70570-0

Library of Congress Cataloging-in-Publication Data. Main entry under title: Oncogenes and human cancer, blood groups in cancer, copper and inflammation, human insulin. (Progress in clinical biochemistry and medicine ; 2) 1. Cancer-Addresses, essays, lectures. 2. Inflammation-Addresses, essays, lectures. 3. Copper-Physiological effect-Addresses, essays, lectures. 4. Insulin-Addresses, essays, lectures. 5. Oncogenes-Addresses, essays, lectures. I. Boehm, T. L. J. (Thomas L. J.), 1956–. II Series. [DNLM: 1. Blood groups. 2. Copper-blood. 3. Inflammation. 4. Insulin. 5. Neoplasms-etiology. 6. Oncogenes. W1 PR668EM v.2/QZ 202 0543] RC265.O63. 1985. 616.99'4. 85-14687.

Typesetting and printing: Schwetzinger Verlagsdruckerei GmbH, 6830 Schwetzingen, Germany
Bookbinding: J. Schäffer OHG, 6718 Grünstadt, Germany
2152/3140-543210

Editorial Bord

Table of Contents

Oncogenes and the Genetic Dissection of Human Cancer: Implications for Basic Research and Clinical Medicine*

Thomas L. J. Boehm

Zentrum der Kinderheilkunde und Zentrum der Biologischen Chemie, Johann Wolfgang Goethe-Universität Frankfurt am Main, Theodor Stern-Kai 7, D-6000 Frankfurt am Main 70, FRG

In recent years, considerable progress has been made in the identification of cellular genes implicated in the genesis of human neoplasia. Among the first evidence for dominant transforming genes were transfection experiments, in which DNA-mediated gene transfer was used to pass from tumor cells a malignant phenotype onto normal fibroblasts. Some of these genes were found to be cellular homologues of viral oncogenes, which confer the ability to cause cancer to acutely transforming retroviruses. With the advent of banding techniques, cytogenetic studies have revealed characteristic abnormalities in several forms of human cancer. To date, the correlation between these non-random karyotypic abnormalities and the chromosomal localization of cellular oncogenes is of limited value. Further, only in very few cases, a cellular oncogene can be directly implicated in the genesis of human tumors. Among the notable advances, however, are findings relating the function of cellular oncogene products to cell proliferation control, since cellular oncogenes code for growth factors, growth factor receptors, and mediators of intracellular message systems. In addition, it was possible to show that the cooperation of oncogenes might by a fundamental principle in oncogenesis, and the genetic dissection of tumorigenesis in at least two parts, immortalization and transformation, provides a frame-work for further studies. Of the alterations of cellular oncogenes in human tumors by amplification, base-pair mutation, and rearrangement, respectively, neither mechanism appears to be sufficient to trigger oncogenic transformation. However, these alterations can serve as definite markers for the classification of human tumors. This concept is elaborated for T- and B-cell leukemia/lymphoma, chronic myelogenous leukemia, Burkitt lymphoma, and retinoblastoma.

* Dedicated to my wife

Progress in Clinical Biochemistry
and Medicine, Vol. 2

Abbreviations

DNA	deoxyribonucleic acid
cDNA	complementary deoxyribonucleic acid
RNA	ribonucleic acid
mRNA	messenger ribonucleic acid
AMP	adenosine monophosphate
GTP	guanosine triphosphate
arg	arginine
cys	cysteine
gly	glycine
glu	glutamine
leu	leucine
val	valine
DMSO	dimethyl sulfoxide
G-CSF	granulocyte colony stimulating factor
ALV	avian leukosis virus
AbMuLV	Abelson murine leukemia virus
HTLV	human T-cell leukemia virus
LLV	lymphoid leukosis virus
HSV	herpes simplex virus
MMTV	mouse mammary tumor virus
BEV	baboon endogeneous virus
RSV	Rous sarcoma virus
SV 40	simian virus 40
LTR	long terminal redundancy
Ad 2 E 1 A	adenovirus type 2 early 1 A region
Ph′	philadelphia (marker) chromosome
HSR	homogeneously staining regions
DM	double minutes
ABR	abnormally banded regions
ALL	acute lymphocytic leukemia
AML	acute myelogenous leukemia
BL	Burkitt's lymphoma
CLL	chronic lymphocytic leukemia
CML	chronic myelogenous leukemia
EGF	epidermal growth factor
PDGF	platelet-derived growth factor

1 Introduction

At the beginning of this century, Boveri suggested that specific genetic alterations may be the basis of tumorigenesis in both animals and man[1]. Since then, explicit genetic determinants of tumorigenesis were indeed defined in such diverse organisms as plants and man. The recent years have witnessed a dramatic advance of knowledge in tumor biology and the convergence of several routes of inquiry leading to a new concept of the genetics of human cancer. In the present article, I will elaborate this concept and will then discuss its applications to clinical oncology.

The first three chapters will discuss different approaches to identify cellular genes implicated in the genesis of human neoplasia (transfection assays, tumorigenesis by retroviruses, cytogenetics). In the following chapter, the role of these specific cellular loci (operationally defined as cellular oncogenes) in the multistep origin of human neoplasia is elaborated with respect to their expression in normal and malignant cells, the mechanisms of alterations in cellular oncogenes in human tumors, and the function of normal and mutated gene products. A tentative model for the cooperation of cellular oncogenes is presented in a concluding paragraph. A final chapter is concerned with the genetics of specific diseases with emphasis on their use in clinical diagnosis and management.

2 Identification of Cellular Genes Implicated in the Genesis of Human Neoplasia

2.1 *Transforming Activities of Tumor DNA*

2.1.1 *Assay Systems*

The technique of DNA-mediated gene transfer was used to analyse whether normal cells could take up DNAs from tumors and thereby become transformed. The analyses of dominant acting tumor genes by transfection were performed with an immortal mouse fibroblast cell line, NIH 3T3, which appeared to take up and express tumor specific DNAs. Early studies investigated the transmissibility of the transformed phenotype from cell to cell via purified DNA using the calcium-phosphate transfection technique developed by Graham and van der Eb[2], and demonstrated the induction of morphological changes as the result of expression of dominant acting tumor genes in chemically transformed cell lines[3,4] and spontaneously occuring tumors[5]. The NIH 3T3 system is extensively reviewed in Refs. 6–8. A non-tumorigenic chinese hamster embryo fibroblast cell line (CHEF/18) was also used to detect transforming genes[9], as well as human cells[10]. The identification of transformed foci in the NIH 3T3 assay is sometimes difficult and depends on cell culture conditions. Therefore, an alternative screening method has been developed by Blair et al.[11]. They injected freshly transfected NIH 3T3 cells into athymic nude mice and scored the induction of tumors as a measure of transformation. In this way, they could detect transforming genes in the DNA of tumor cell lines which were not readily observed in the conventional NIH 3T3 assay.

2.1.2 Survey of Tumors for Transforming Genes

The NIH 3T3 transfection assay for the detection of dominant acting transforming genes yields positive results only in a fraction of tumor DNAs tested, about 20–30%. This result may either mean that the negative tumors do not possess dominant genes or that the NIH 3T3 assay is restricted to only a portion of such sequences. Nevertheless, the NIH 3T3 system has allowed the identification and, in some cases, the molecular cloning of such dominant acting genes. Initially, the transfected transforming genes were classified according to their sensitivity of prior cleavage with restriction enzymes. In such experiments, the tumor DNA was cleaved with particular restriction enzymes and then tested for transforming capability. Such experiments allowed the identification of several distinct genes, suggesting that tumor-derived cell lines contained both common and distinct transforming elements[12)] and that related transforming genes are activated in the same type of tumor even from different species[13)]. Further, stage-specific transforming genes of human and mouse B- and T-lymphocyte neoplasms were detected[14)]. Among the genes detected in the NIH 3T3 system were members of the *ras* family of oncogenes (see 3.2.2), as well as oncogenes unrelated to this family (Blym-1, Tlym-1, the transforming gene from the human MNNG-HOS osteosarcoma cell line, and that found in colon carcinomas 1665 and 2033[15–18)]). Further, a c-sis cDNA clone (pSM-1) isolated from the human HUT-102 cell line (derived from a cutaneous T-cell lymphoma and infected with HTLV) led to the transformation of NIH 3T3 cells[19)]. Clearly, a number of transforming genes must exist in addition to those detected by the NIH 3T3 assay: however, with all its limitations, this assay procedure provided a wealth of new information to the characterization of transforming genes in human tumor DNA.

2.2 A Model: Viruses that Cause Cancer

The induction of animal tumors by viruses has long been established. It was only in recent years that this research fostered a new understanding of the genetics of human cancer, after it was found that the viral oncogenes found in acutely transforming retroviruses were in fact derived from evolutionary conserved vertebrate genes. Further, the understanding how non-acutely transforming retroviruses cause tumors in infected animals suggested a new mode of tumorigenic events also in human tumors. The following chapter thus concentrates on the illustration of these different ways of action, which are indirectly or directly dependent on the genetic information provided by the host cell.

2.2.1 *Insertion of a Retrovirus Adjacent to a Cellular Oncogene can Lead to Tumorigenesis*

This mechanism is the prime path for the so-called non-acutely transforming retroviruses, which, while carrying no oncogene, can induce, with long latency, tumors in infected animals. The case of avian leukosis virus (ALV) is the best studied. ALV can induce lymphoid leukosis and erythroblastosis, as well as fibrosarcoma, nephroblastoma, and hemangiosarcoma[20]. It is tempting to speculate that each tumor type is induced by alteration of a specific cellular locus by the integration of ALV provirus. It was found that ALV integrates adjacent to c-myc in most cases of lymphoid leukosis[21, 22]. This integration was shown to disrupt the normal pattern of DNAaseI hypersensitive sites that lie upstream from c-myc in normal chicken cells, yielding a long terminal repeat (LTR)-associated site, thus representing a more efficient entry site for various transcriptional factors than the endogeneous c-myc hypersensitive sites[23]. In ALV-induced erythroblastosis, integration of the provirus was detected adjacent to the c-erbB locus[24]. Thus, the oncogenic potential of slowly oncogenic retroviruses resides in their ability to mutagenise or activate cellular loci; their oncogenic spectrum may then, at least in part, not only depend upon the type of cells they infect, but also upon the genes they activate. The progression of such tumors may, in addition, involve proviral deletions and oncogene base-substitutions in insertionally mutagenized cellular oncogene alleles, as shown for the c-myc gene in ALV-induced bursal lymphomas[25]. This finding explains why the so-called promotor interference[26] *) is not found in naturally occuring tumors, since, very often, the upstream promotor of the integrated provirus is deleted or damaged to allow the 3′-LTR to function as an efficient promotor.

An analogous situation to the avian lymphoid leukosis induced by ALV can be found in the case of the murine leukemia virus (MuLV), which integrates adjacent to c-myc in mouse lymphomas[27]. Integration into the so-called Pim-1 region was found in early T-cell lymphomas[28]. Mouse mammary tumor virus (MMTV) integrates into the int-1 region and this is accompanied by the appearance of transcriptional products from int-1, a silent domain in non-neoplastic mammary tissue[29]. The stimulation of transcription is, however, indirect and may be due to a cis-acting enhancer function of MMTV[29]. MMTV can also integrate into a different cellular locus, called int-2[30]; it is not known, whether these two sites represent distinct cellular genes, or whether they function independently or as parts of a single biochemical pathway. They are, at least, located on different chromosomes[30a].

The human T-cell lymphoma virus (HLTLV-1) is an interesting case for the situation in man. HTLV is able to transform normal cells in culture to a state of unlimited proliferation potential, the exact mechanism for this "immortalization" predominantly of a subset of T-cells of the OKT 4+ phenotype being obscure as yet. The absence for a common region for provirus integration in leukemic cells suggests that it is unlikely that this disease is caused by a simple insertional mutagenesis; since

*) Viral mRNA transcription extending to the downstream LTR of a provirus-like construct restricts the ability of this LTR to function as an efficient promotor. This inhibition is removed if the upstream LTR is damaged or deleted or if a transcriptional termination signal is inserted between the LTRs[26].

integration also occurs on different chromosomes one might suggest that there is a genetic structure inherent in HTLV which is responsible for its effects[31, 33]. By molecular analyses, it was shown that all HTLV viruses (group I and group II) share a common sequence, called pX, which codes, at least by its overlapping reading frames, for two different peptides, none of which has been identified so far in transformed cells[33a]. These proteins are, however, believed to play a role as transcriptional activators of the LTR and, possibly, other cellular control elements. In the latter respect, it is interesting that in HTLV-1 infected cells, a cellular gene is abundantly expressed. This gene, specified by its cDNA clone HT-3 presumably represents a gene important in T-cell proliferation (a growth factor or growth factor receptor etc.)[34].

The fact that the infection of cultured human cells with baboon endogeneous virus (BEV) frequently leads to an association of viral DNA with a specific locus (termed BEVI) suggests the possibility of targeted integration of a virus into a host genome[35]. This phenomenon, however, may also be explained by a selection mechanism believed to occur in all retrovirus-genome interactions and mimicking a nonrandom integration simply by confering a growth or other advantage to cells with a specific integration site.

Another interesting example for insertional mutagenesis is the case of c-mos in the mouse. The c-mos locus is transcriptionally silent but is rearranged in a mouse myeloma and the tumor RNAs contain a transcript hybridizing with a viral mos probe[36]. The c-mos gene is activated by the insertion of an endogenous intracisternal A-particle genome[37, 38] and the orientation of the integrated IAP relative to c-mos is correlated to the degree of activation of c-mos (a tail-to-head configuration is less active than a head-to-head configuration)[39].

Taken together, the data reviewed in this paragraph suggest that random insertion of a retroviral genome can, if it happens to invade a cellular oncogene locus, lead to the selection of tumorigenic variants by inappropriate activation of a cellular gene.

2.2.2 Insertion of a Cellular Oncogene Between the Controlling Sequences of a Retrovirus Leads to an Oncogenic Retrovirus

Non-transforming retroviruses are, in simplified terms, composed of four genetic elements, long terminal redundancies (LTR) containing enhancer and promotor sequences at both ends of the genome, a coding region for the viral core proteins (gag), the pol gene for the RNA-dependent DNA polymerase (reverse transcriptase), and the env complex coding for the viral envelope proteins. Non-acutely transforming retroviruses, which exert their oncogenic potential only after long latency by mechanisms discussed above, have such a genetic constitution. In contrast, most acutely transforming retroviruses have suffered alterations in this genetic building block by insertion of new genetic material, derived from host cellular sequences. The insertion can be of various types (see Ref. 40 for review); for example, in the Rous sarcoma virus (RSV), the new genetic material, termed src, is located 3′ to the env gene and thus preserves the replication competence of the virus. In the other known retroviruses, where the recombinational event disrupts vital parts of the retrovirus, the resulting viruses are replication defective, that is, they are dependent on the presence of a replication competent helper virus in order to integrate into the host genome. Of special interest is the occurrence of retroviruses which contain two different oncogenes, exemplified by the avian erythroblastosis virus (AEV), the avian leukemia virus E 26, and the murine leukemia virus MH 2. Their genetic composition is a follows: AEV, LTR-Δ gag-erbA-erbB-Δ env-LTR[41], E 26, LTR-Δ gag-myb-ets-Δ env-LTR[42, 43], and MH 2, LTR-Δ gag-mil-myc-LTR[44–48]. The erbA and erbB genes, and the myc and mil genes are transcribed from different transcriptional units, whereas the myb and ets genes are comprised in a large mRNA and are expressed as a triple fusion protein, $p^{135\Delta\ \text{gag-myb-ets}}$. A more subtle oncogene hybrid may be represented by the v-fgr gene, which has homology both with actin, a eukaryotic cytoskeletal protein, and with the tyrosine-specific kinase gene family[48a]. The natural ocurrence of such viruses indicates either two successive transducing events in the generation of such viruses or the possibility that a non-random translocation in the chicken/mouse genome had brought together these two sequences prior to being taken up by the retrovirus. It is of interest in this respect that the cellular loci for erbA and erbB are on different chromosomes in chicken[49] and man[50], but on the same in the mouse[50].

An important question in the genesis of acutely transforming retroviruses is the relationship between the cellular and the viral oncogene. There is no example where v-onc is isogenic to c-onc, suggesting that the cellular progenitors have to be modified in addition to being controlled by viral regulatory mechanisms to exert a truely oncogenic effect. This hypothesis was tested directly for the c-src and the c-fos genes. Iba et al.[51] examined the effects of overproduction of $p60^{c-src}$ in chicken embryo fibroblasts, the natural host of RSV, by introducing RSV variants that encode $p60^{c-src}$ or recombinant proteins between $p60^{c-src}$ and $p60^{v-src}$. RSV variants that encode c-src expressed a large amount of $p60^{c-src}$ in infected cells but failed to cause cell transformation, indicating a qualitative difference between c-src and v-src. After transduction of the c-src gene into a retrovirus, this gene is subject to a high rate of mutation and transforming variants can readily be obtained. The c-fos gene can lead to cell transformation after being linked to a viral promotor sequence[52], although c-fos and v-fos differ at sequences at the C-terminus, probably, as a result of an out-of-frame deletion[53]. Another informative example is the relationship between c-abl and v-abl; v-abl is an internal fragment of the c-abl mRNA, and there is at least one point mutation which results in an amino acid substitution[54].

The fact that cellular and viral oncogenes are not isogenic, although they share specific sequences and differ from each other in scattered point mutations and unique coding regions, led Duesberg[55] to conclude that the oncogenic potential of cellular oncogenes is reflected as progenitors of viral oncogenes and that cellular oncogenes can participate in tumorigenesis only when qualitatively altered.

A model for the transduction of cellular genes by retroviruses has been advanced by Bishop et al.[56]. The transducing vehicle is a preexistent retrovirus, whose proviruses happen to integrate upstream from a cellular oncogene. The important event, mediated by DNA rearrangement of possibly two different modes, is the formation of a hybrid transcriptional unit. In the first possibility, the rearrangement transposes the oncogene into a position within the provirus, transcription and splicing would now produce a possibly defective but otherwise conventional retrovirus bearing at least part of the cellular oncogene. The second possibility, thought to occur more often, fuses provirus and oncogene together into a single genetic element by a deletion event. The viral promotor is then joined by a variable part of the cellular oncogene including sites for the termination of transcription, splicing etc. In this case, transcription begins at the viral promotor, proceeds through the cellular oncogene sequence, and generates a hybrid RNA from which the introns are removed by splicing. Recombination can then occur when hetereozygous virirons (that is, a heterodimer between recombinant and wild-type retrovirus) enter a new host cell and begin reverse transcription.

When one analyses the kind of genes which are transduced by oncogenic retroviruses, there is only a limited number of genes, which are often transduced by different viruses. This suggests that only a small number of cellular genes is capable, after suitable modification (by deletion, mutation, etc.) to mediate malignant growth when controlled by the viral promotor activities. It has furthermore been noted that some of these genes belong to gene families, and that some represent single, unrelated isolates. The largest family is the so-called src-family, named after the oncogene of Rous sarcoma virus, which was identified first as being a transduced oncogene. The evolutionary relationship among the members of the src-family (comprised of mos, mil(mht)/raf, erbB, fms, rel, yes, src, abl, fes/fps, fgr, and ros) has been extensively discussed[57)]. It is worth noting here that several evolutionary conserved cellular genes have been transduced by retroviruses from different species, for example, mil(mht) and raf represent the avian and murine, respectively, homologues of a single cellular oncogene[46–48, 58)]. The same is true for fes and fps[59)]. The viral oncogenes yes, src, abl, fes/fps specify a tyrosine-specific kinase activity which has distant relationship to the catalytic chain of the bovine cyclic AMP-dependent protein kinase[60)]. The viral oncogenes mos, mil(mht)/raf, and erbB have sequence homology to v-src, but do not code for a tyrosine-specific kinase activity. The same is true for fms and rel, for which the exact evolutionary position in the src-family has not been established. fgr and ros also code for tyrosine kinase but their position in the evolutionary tree is uncertain as well. It should be noted that sequences homologous to src and abl have been found in the DNA of Drosophila[61)] and were shown to be developmentally regulated[62)], suggesting that these genes function at crucial points in cellular differentiation and growth in eukaryotic organisms.

Another large family is the so-called ras-family, consisting, in man, of c-ras-H, c-ras-K, and c-ras-N. The former two have been found transduced in several retroviruses, whereas the latter was identified so far only in transfection experiments. It is of interest that ras-related sequences were also found in yeast[63–66)].

The myc-family comprises of the c-myc locus transduced into several avian and feline retroviruses, and the c-myc-N gene, detected by hybridization survey of neuroblastoma DNA (see 3.2.1).

2.3 From Chromosomes to Genes: Refining the Genetics of Human Cancer

For a long time, the complexity and profound heterogeneity of karyotypic findings in human neoplasia precluded the definition and characterization of specific cytogenetic lesions in particular disease entities. With the introduction of banding techniques, however, this situation gradually changed and as of to date, a number specific chromosomal changes associated with neoplastic conditions have been found. These non-random karyotypic changes are the basis for the chromosome hypothesis of oncogenesis, founded by Boveri in 1914[1)] and recently advanced primarily by Yunis and Sandberg[67, 68)]. In their view, a certain karyotypic change must occur and is responsible for the malignant cellular manifestation. It is now generally accepted that chromosomal defects are the rule rather than the exception in human neoplasia and it seems that the cytogenetic findings have suggested to molecular biologists the regions of the chromosomes where to concentrate on. As a result of this cooperation, the recent years have wittnessed a dramatic refinement in the genetic analysis of human tumor cells. In this chapter, we will compile the present data relating non-random karyotypic changes in cancer and leukemia to the locations of specific cellular genes thought to be important in oncogenesis.

2.3.1 Characteristic Abnormalities of Chromosomes in Human Cancer

Karyotypic changes can be grouped into three distinct classes[69)]. The first class represents reciprocal translocations in which there is no loss of structural material. The second class comprises of deletions and non-reciprocal rearrangements resulting in the loss of structural material at these sites. The duplication of whole chromosomes and their segments (i.e. gene amplification, double minute chromosomes, homogeneously staining regions) constitutes the third class of chromosomal aberrations. It is commonly held that there are two different steps in the evolution of malignant disease with respect to karyotypic abnormalities. Defects present early in the development of neoplasia are generally considered to be related to the etiology of the specific disease, whereas other recurrent defects (usually apparent only in the progression of the disease) are viewed as contributing to the tumor heterogeneity, although the selection of cells with complex aberrations cannot be excluded. The primary karyotypic changes involve genes controlling either differentiation and proliferation, or both, whose absence or altered expression confers the ability to the cell to become malignant. Such, the exchange of material between chromosomes could contribute to the activation or silencing of genes carried by these segments in several ways, which are not mutually exclusive: (i) the gene in question moves away from an adjacent cis-acting activator/suppressor, (ii) the gene moves next to an endogenous promotor region, and (iii) the gene structure is broken by the translocation process thereby altering the provision of a correct mRNA molecule.

In Table 1, several non-random cytogenetic abnormalities have been compiled. It is apparent that only a limited number of chromosomal regions is involved in human cancer and leukemia, and that several distinct types of disease share part of their karyotypic abnormalities.

Table 1. Correlation between non-random karyotypic alterations and associated neoplastic diseases with chromosomal localizations of fragile sites and cellular oncogenes[a]

chr. no.	karyotypic alteration	breakpoint or deletion	associated neoplasm	fragile site	oncogene location		Ref.
1	del 1 p		malignant melanoma		1 p 32	c-B-lym 1	77
	del 1 p	1 p 32 → 1 pter 1 p 31 → 1 p 36	neuroblastoma		1 p 32 → cen 1 p 31.1 → p 22.1 1 p 11 → p 13	c-N-ras	78 79 80
	t (1;4)	1 q 23 and 4 q 32	lymphoma		1 p 34 1 q 12 → qter	c-src-1 c-sk-1	97 79
2					2 p 12 2 p 23/24	Ig κ c-N-myc	81 82
	t (2;8)	2 p 12 and 8 q 23	Burkitt's lymphoma	2 q 13			
3	del 3 p	3 p 14 → p 23	small cell cancer of lung	3 p 14	3 p 25	c-raf-1	83
	t (3;8)	3 p 25 and 8 q 21	mixed tumor of parotid				
4	t (4;11)	4 q 21 and 11 q 23	acute lymphocytic leukemia		4	c-raf-2	83
5	del 5 q	5 q 22 → 5 q 23	ANLL (M 1, M 2, M 4, M 5, M 6)		5 q 34	c-fms	79, 84
6	i (6 p)		malignant melanoma	6 p 23	6 pter → q 13 6 p 23 → q 12	c-ki-ras-1	85 86
	del 6 q		malignant melanoma		6 q 15 → q 21	c-myb	85
	t (6;14)	6 q 21 and 14 q 24	serous cystadenocarcinoma of ovary		6 q 22 → q 24 6 q 21 → qter		87 87 a
7	+7		adenocarcinoma of large bowel	7 p 11			
	del 7 q	7 q 33 → 7 q 36	ANLL (M 1, M 2, M 4, M 5, M 6)		7 pter → 7 q 22	c-erb-B	88
8	+8		ANLL (M 1, M 2, M 4, M 5, M 6) bladder cancer				
	t (8;21)	8 q 22.1 and 21 q 22.3	AML (M 2), ANLL (M 2)	8 q 22	8 q 22 8 q 24	c-mos c-myc	89 89
	t (8;14)	8 q 24.13 and 14 q 32.33	Burkitt's lymphoma, ALL (L 3)				

	t (8;22)	8 q 24 and 22 q 11.23	Burkitt's lymphoma				
9	t (9;11)	9 p 22 and 11 q 23	AML	9 p 21			
	del 9 p		T-cell ALL				
	t (9;22)	9 q 34.1 and 22 q 11.21	CML, AML(M1), ALL (L1, L2), ANLL (M1)	9 q 32	9 q 34.1	c-abl	74, 90
10				10 q 23 10 q 25			
11	del 11 p	11 p 13	Wilms' tumor		11 p 11→11 p 15	c-Ha-ras-1	91
	del 11 q		ANLL				
	t (11;14)	11 q 13 and 14 q 32	CLL	11 q 13 11 q 23	11 q 23→q 24	c-ets	92
	t (11;21)	11 q 22 and 21 q 21	myeloproliferative disorders and ANLL				
	t (11;22)	11 q 23/24 and 22 q 12	Ewing's sarcoma				
12	+12		CLL				
	i (12 p)		seminoma				
	del 12 q		adenocarcinoma of large bowel	12 q 13	12 q 12→pter	c-Ki-ras-2	86
13	del 13 q	13 q 14.13	retinoblastoma				
14	14 q+		lymphoma, ALL, CLL, adult T-cell leukemia				
	t (14;14)	14 q 24 and 14 q 32	lymphoma		14 q 32	IgH	93–95
	t (14;18)	14 q 32.3 and 18 q 21.3	lymphoma				
15	t (15;17)	15 q 22 and 17 q 11.2	ANLL (M 3)		15 q 26.1 15 q 25→q 26	c-fes	90 87
16	inv (16)	16 p 13.2 and 16 q 22	ANLL (M 4)	16 p 12 16 q 22			
	del 16 q		ANLL				

Table 1 (continued)

chr. no.	karyotypic alteration	breakpoint or deletion	associated neoplasm	fragile site	oncogene location		Ref.
17	i(17 q)		CML-blast crisis	17 p 12	17 p 11→q 21 17 q 21→q 24	} c-erb-A-1	88 96
18							
19							
20	del 20 q		polycythemia vera	20 p 11	20 q 12–13	c-src-2	97
21	+21 del 21 q		ALL ANLL				
22	del 22 q		CML, meningeoma		22 q 11 22 q 12.3→q 13.1	Ig λ c-sis	98, 99 76
X				Xq 27	c-Ha-ras-2		100

[a] The list of karyotypic alterations has been compiled from Refs. 67 and 68, the list of fragile sites is taken from Ref. 70. The studies for the assignment of oncogenes are referenced individually; also given are the locations of the immunoglobulin gene complex, since this is relevant to the discussion of Burkitt's lymphoma. The abbrevations of cellular oncogenes follow a published system[66a)]

2.3.2 Fragile Sites in Human Chromosomes

Fragile sites in human chromosomes are identified as chromosomal regions in the human genome which fail to become stained in routine banding procedures and at which breakage of chromosomes can be observed in such analyses. These sites are most easily identified under special culture conditions (i.e. depletion of cells of folic acid and thymidine). It is remarkable that these sites are inherited in a truely Mendelian fashion so that same locus is involved in any particular family. Of 17 heritable fragile sites identified, 16 are located on autosomes[70]. Only one is located on the X chromosome (Xq 27) and is a marker of one form of the X-linked mental retardation syndromes. The incidence of fragile sites in the human population is not known, although current estimates are far less than 1 of 1000. Table 1 also lists the locations of these sites in the human chromosomal map and indicates that 7 out of 17 fragile sites occur in chromosomal bands involved in non-random abnormalities and that 12 of 13 chromosomes that contain fragile sites are involved in non-random structural and numerical changes in malignant cells. If the coincidence of fragile sites at the localization of chromosomal aberrations is meaningful, one might expect cancer patients with chromosomal rearrangements also to be carriers of fragile sites in the affected chromosomes. Data reported by Yunis[71, 72] and Rowley[70] suggest that this might indeed be the case. It is not clear, however, whether such persons do have fragile sites also in their normal cells and what the mechanism is for the induction of chromosomal defects at fragile sites. Fragile sites may be prone to breakage and this may prove to be a predisposing factor in the development of specific neoplastic diseases.

2.3.3 Chromosomal Localization of Cellular Oncogenes

It has already been discussed that cellular oncogenes are suspected of key functions not only in normal differentiation and development but also in maligancy. It is of interest therefore to compare the chromosomal positions of these genes with the sites of non-random karyotypic abnormalities. A number of different techniques were used to localize the cellular oncogenes to specific chromosomal positions, among which the analysis of somatic cell hybrids and the in situ hybridization techniques were the most successful ones.

If one analyses the correlations depicted in Table 1, it is clear that a number of cellular oncogenes can be implicated in the genesis of specific neoplastic diseases. The most striking example is the case of chronic myelogenous leukemia (CML) in which the reciprocal translocation between chromosomes 9 and 22[73] occurs exactly at the same position where the c-abl gene resides. Indeed, it has been shown that this gene is, in Philadelphia chromosome positive cases, translocated to chromosome 22; the c-sis gene, located on the long arm of chromosome 22 is then transferred to chromosome 9[74–76]. This finding, obtained by molecular cloning experiments, not only proves the reciprocal character of this translocation, it also clearly demonstrates that cellular oncogenes can be transferred in a consistent manner to new environments. Work from several laboratories (to be discussed in detail in 4.2) has shown that the breakpoint is not directly in the c-abl gene itself, but is confined to a region outside the coding regions ([74, 75], our unpublished observations). The second striking example is the case of Burkitt's lymphoma which was long recognized because of its three characteristic translocations, involving c-myc on chromosome 8 and the immunoglobulin gene loci on chromosomes 2, 14, and 22. These studies are extensively treated in 4.3.

By comparing the chromosomal assignment of cellular oncogenes with the breakpoints listed in Table 1, it is evident that one can suspect such a relationship for only some of these genes. Experimental proof for this assumption is, in the majority of cases, still lacking. Correlations between location of fragile sites and cellular oncogenes are of limited value either. Although 10 of the 11 autosomes carry both an oncogene and a fragile site, only in one case, their locations are coincident, namely at 8q22 which contains the c-mos gene. The fact that this locus may represent a pseudogene, does not strengthen the notion that fragile sites may hint the presence of oncogenes.

To conclude, it is clear that karyotypic analyses identify a number of non-random abnormalities of human chromosomes in neoplastic cells. In numerous cases, characteristic lesions could be identified. This may contribute not only to the classification of diseases but also to the detection of biologically distinct subgroups. The refinements offered by molecular biology techniques, notably the advances offered by recombinant DNA technology, and the isolation and characterization of cellular oncogenes can greatly contribute to the genetic characterization of neoplastic diseases (see 4.).

3 A Multi-Step Origin of Human Neoplasia: Are Oncogenes Involved?

3.1 Expression of Oncogenes

Early studies using rodent species have shown that cellular oncogenes might function in the regulation of cellular growth and differentiation in vertebrate development. Müller et al.[115] studied the pre- and postnatal development of the mouse with respect to the expression of several oncogenes. The expression of c-fos probably occurs in cells which give rise to the placenta and in cells of mesenchymal origin. The expression of c-ras-H is almost the same in all tissues examined and at all developmental stages. The highest level of c-abl expression is found in the testes of young adult mice and also in lymphatic tissues. The analysis of the complex pattern of transcripts of sizes 3.7, 4.7, and 5.7 kb, respectively, revealed variable amounts of individual transcripts, suggesting that initiation and/or termination of c-abl transcription, and/or splicing of the c-abl related transcripts may be a strictly controlled process. This study demonstrated clearly that the expression of cellular oncogenes is both tissue-specific and dependent on the developmental stage. This may also mean that the expression of these genes can be a marker of cellular differentiation or indicative of different growth rates, or both. This suggestion is supported by studies on the expression of c-myb in hematopoetic tissues of mice, which appears to be developmentally regulated and important during T-cell differentiation[101]. Other studies have specifically asked whether some oncogenes might be associated with a proliferative state of cells, whereas others might be markers of differentiation. This is most easily demonstrated in systems where cells can be made to differentiate upon an exogeneous stimulus. Gonda and Metcalf[102] have studied the expression of oncogenes in the murine myelomonocytic leukemia line WEHI-3B, which upon addition of G-CSF or actinomycin undergoes differentiation yielding primarily monocytic cells. The expression of c-ras-H was unchanged, the expression of c-myb and c-myc declined 10-fold, whereas the expression of c-fos increased about 11-fold. This indicates that c-myb and c-myc do not themselves control myeloid differentiation (c-fos is a likely candidate for macrophage-specific functions), but that they function in the maintenance of the proliferative state of myeloid cells. Westin et al.[103] studied the expression of c-myb in a variety of established cell lines and detected its expression in all immature myeloid and lymphoid T-cells, in addition to an erythroid cell line. Mature T and B cells did not express this gene. They also studied the expression of c-myb during DMSO or retinoic acid stimulated differentiation of HL-60 cells. With the appearance of mature myeloid cells, c-myb expression declined. In a parallel study, these authors showed that also the c-myc gene but not c-abl expression was declining during HL-60 differentiation[104]. The human myeloblastic leukemia cell line ML-1 can be induced to differentiate into monocyte and macrophage-like cells by the tumor promotor TPA. This is accompanied by a 5-fold reduction of c-myb expression *prior* to the loss of DNA synthesis that accompanies the differentiation process[105].

These data indicate that expression of c-myb is correlated to the proliferative state in myeloid cells. A clinical study supports this notion. Rosson and Tereba[106], while studying the levels of expression of various oncogenes in childhood leukemia noted that the c-myc gene was expressed in all samples tested, whereas c-myb expression was primarily associated with immature myeloblasts. The c-ras-H gene was expressed in all samples tested, albeit at low levels.

The expression of c-myc has been studied with respect to the cell cycle. In liver regeneration, changes in c-myc expression occur before DNA synthesis, suggesting a role in triggering the S phase in hepatocytes[107–109]. The in vitro stimulation of B lymphocytes and fibroblasts is also known to induce an immediate increase in the expression of c-myc[110]. This increase is temporary, and the expression returns to uninduced levels by the time the DNA replication starts. The human lymphoblastoid cell line Daudi is sensitive to growth inhibition by human β-interferon. This is accompanied by a selective reduction of c-myc mRNA levels (the level of actin mRNA served as a control)[112]. The human promyelocytic cell line HL-60 also responds to dihydroxy metabolite of vitamine D_3 by ceasing division and aquiring some of the characteristics of either granulocytes or monocytes. These phenotypic changes are preceeded by a marked decrement in the expression of the c-myc gene[113]. It was also found that in nontransformed mouse cells, c-myc could be induced by growth stimulation, but that its expression was constitutive in two chemically transformed derivative cell lines[111]. Thus, altered regulation of c-myc expression might prevent cells from entering G_0 phase and thus lead to their infinite growth.

Table 2. Expression of cellular oncogenes in human cells and tissues[a]

c-onc	cell or tissue examined		size of transcript (kb)	Ref.
	normal cells/tissue	tumor cells/tissue		
c-fos[c]	amnion chorion	vulva carcinoma cell line	2.2[b]	115
		vulva carcinoma cell line renal cell carcinoma (7/9)[d] ovarian adenocarcinoma (6/6) colon adenocarcinoma (5/5) rectal adenocarcinoma (3/3) carcinoma of lung (4/4) carcinoma of breast (4/4) hematologic malignancies (12/14)		114
c-fms	placenta[e 115)]	renal cell carcinoma (4/9) ovarian adenocarcinoma (4/6) carcinoma of lung (2/4) carcinoma of breast (3/4)	3.6	114
		LICR myeloma cell line DC 199 melanoma cell line		116
		choriocarcinoma (trophoblast-like)	3.7[b]	115
c-myb	peripheral blood myeloblasts	promyelocytic cell line HL-60 immature myeloid cell line KG-1 erythroid precursor cell line K 562 immature T-cell lines CCRF-CEM Molt 4 KM-3 T-cell ALL blasts	4.5	104
		myeloblastic leukemia cells ML-1	4.3	105
		various hematologic malignancies	3.4	114
c-fes		cancer of lung (4/4) carcinoma of breast (3/4) ALL (7/7) AML (2/2) CML (2/2)	2.6	114

[a] Expression of c-myc, c-ras-H, c-ras-K was found in all cells and tissues so far examined; the expression of c-myc will be discussed in section 4.3. The size of transcripts from the c-ras-K locus is 4.6 kb, that from the c-ras-H locus about 1.4 kb[114)]. The latter is found as similarly sized transcripts in tissues of the mouse[115)], and the rat[108)]. The expression of c-abl in human tissues is not yet fully understood, primarily for technical reasons. Workers using a DNA probe encompassing the kinase domain have found c-abl related transcripts in all samples studied. Since DNA sequences coding for the kinase domain may also cross-hybridize with cellular tyrosine-phosphokinase mRNAs, this may not be a true reflection of c-abl expression. Accordingly, Slamon et al.[114)], while using a probe not containing kinase coding sequences have detected c-abl expression in only 3 of 54 tumors tested. Probes derived from the human c-abl locus will be needed to clarify this situation

[b] A transcript of similar size is also found in mouse tissues[115)]

[c] Low levels of expression were found in all human tissues tested[118)]

[d] Number of positive samples/total number tested

[e] Low levels of expression were also found in fetal membranes[115)]

Table 2 (continued)

c-onc	cell or tissue examined		size of transcript (kb)	Ref.
	normal cells/tissue	tumor cells/tissue		
c-src		CML (1/2) Lymphosarcoma (single case studied)		114[f]
		osteosarcoma (single case studied) Wilms' tumor (single case studied)		106
c-sis		sarcoma cell lines (5/6) glioblastoma cell lines (3/5)	4.2	117
		T-cell line HUT-102 (HTLV-producer)	4.2	103
		melanoma cell line DC199 myeloma cell line LICR		103
c-erb-A		K562 erythroid precursor cell		101

[f] A single patient with ALL found to have high levels of c-src and c-myc expression was also studied in remission; the expression of c-src but not c-myc dropped significantly; this was interpreted to mean that a loss of tumor cells had occured that were replaced by normal leukocytes

In conclusion, one can suggest that there are oncogenes which appear to be directly involved in proliferation control. It further appears that there might be some kind of hierarchy in this control in that there are cell-lineage specific and cell-lineage independent requirements for proliferation. The c-myc gene may represent a gene which is ubiquitously expressed in cell proliferation, whereas the c-myb gene expression may be necessary only for myeloid cells. It will be interesting to study whether induced expression of c-myc can force myeloid cells into cell division and whether c-myb expression is required or not. A large clinical study conducted by Slamon et al.[114)] suggests that any oncogene expressed in a tumor sample is also expressed in the normal surrounding tissue. Sometimes, there is a quantitative difference, but a qualitative change was never observed.

Table 2 attempts to summarize some pertinent studies with respect to the type of onc gene expressed and some characteristics of the transcripts. It is notable that there are some oncogenes for which expression was not proved until now (c-erb-B, c-mos, and c-yes), whereas c-myc, c-ras-H and c-ras-K appear to be always expressed.

3.2 Mechanisms of Alteration of Cellular Oncogenes in Human Tumors

The following section will cite examples for changes in copy number (i.e. gene amplification), mutations in coding regions, and rearrangements of transcriptional units of cellular oncogenes which all occur in human neoplasia. It appears that neither mechanism alone can suffice in malignant conversion of normal cells.

3.2.1 Amplification of Cellular Oncogenes

The amplification of cellular oncogenes (i.e. the increase in copy number of this gene) entails the concomitant increase in specific mRNA levels and elevated amounts of the oncogene product. This may disturb delicate balances built up in a cell in order to control cellular differentiation and/or proliferation processes. Table 3 compiles evidence for amplifications of the c-myc, c-myb, and c-abl oncogenes in various human malignancies. It is worth noting that (i) gene amplification does not seem to be a frequent phenomenon and (ii) many of these tumors contained chromosomal abnormalities characteristic of gene amplification, such as DM, HSR, and ABR, respectively. In the case of the apudoma Colo 320, the site of the amplified c-myc gene was shown by in situ hybridization to be in DM and HSR[119)]. That this may, however, not always be the case is illustrated by the colon carcinomas Colo 201 and Colo 205, which have an amplified c-myb on a marker chromosome, yet contain also DM and HSR[120)]. Amplification of ras oncogenes has not been found in human tumors, but was noted for the c-ras-K gene in a mouse adrenocorticoid tumor cell line Y 1[121)]. Two recent studies relate the presence of amplified oncogenes to the clinical course of malignant disease. The c-myc gene was found to be amplified in lung cancer, and it was suggested that the c-myc amplification played a role in the phenotypic conversion and malignant behaviour of small cell lung cancer cell lines into their variant form[122)]. This conclusion was based on the finding that non-small cell lung cancers contained only low levels of c-myc amplification, whereas small cell lung cancer lines were characterized by intermediate levels of amplification. The biologically more aggressive variant form of small cell lung cancer showed levels of up to 100 and more copies of the c-myc gene. Brodeur et al.[123)] have found that the level of amplification of the c-myc-N oncogene in untreated human neuroblastomas correlates with the advanced disease stage; interestingly, a bimodal extent of amplification was found. Whereas in clinical stages 1 and

Table 3. Amplification of cellular oncogenes in tumor cells

cell line/tumor	amplified oncogene	level of amplification (fold)	karyotypic abnormalities[a]	elevated mRNA levels yes/no	Ref.
acute promyelocytic leukemia HL-60	c-myc	16–32	DM, ABR	yes	126, 127, 132
apudoma (Colo 320)	c-myc	16–64	DM, HSR	yes	119
breast adenocarcinoma SK-BR-3	c-myc	8–16	?	yes	128
acute myelogeneous leukemia	c-myb	5–10	ABR at 6q22–24	yes	129
colon carcinoma (Colo 201/205)	c-myb	10	marker chr.	yes	120
erythroid precursor (K562)	c-abl	4–8	?	yes	130, 131

[a] elongated homogeneously staining regions (HSR); small chromosomal paired chromatin bodies (double minutes, DM) which lack centromers and segregate unpredictably at cell divisions; abnormally banded regions (ABR)

2, none of the 15 tumors showed c-myc-N amplification, 10 of 48 and 12 of 48 patients in stages 3 and 4 had amplification levels of up to 10 and more than 100 copies of the c-myc-N gene, respectively. The amplified gene can be translocated from its normal position at 2p 23–24 to any of several other chromosomes[124]. In a parallel study, the same group of investigators hypothesized that the increased expression of the c-myc-N gene consequent to amplification of DNA may contribute to the malignant progression, rather than to the initiation of neuroblastoma[125]. The fact that c-myc-N is also amplified in retinoblastomas[125] may suggest that c-myc-N amplification accompanies tumorigenesis in neurogenic tissues.

In conclusion, the available data indicate the possibility of oncogene amplification in human tumors, with a possibility of tissue specificity, but also strongly suggest that the amplification is not causative in the genesis of the disease; instead, once it occurs, it may merely provide a selective advantage to cells during the progression of the malignant disease.

3.2.2 *Base-Pair Mutations*

As already mentioned, the ras genes are members of a highly conserved family of vertebrate genes, which were first detected as the oncogenes of the Kirsten and Harvey sarcoma viruses. Mutated ras genes have been detected in human tumor cells by their ability to induce tumorigenic transformation of the immortal mouse fibroblast cell line NIH 3T3 (see 2.1.2). Many of these genes have been characterized quite extensively, and it was found that they all contain mutations in either positions 12 or 61 of the encoded proteins (see Table 4). It is notable from these studies, that different point mutations at the same codon can activate a gene. A recent study shows, however, that mutations at other sites are also capable of inducing transformation of NIH 3T3 cells[133]. Single amino acid substitutions at positions 12, 13, 59, or 63 have an activating potential. Nevertheless, it appears that there are two major hot spots on the ras protein which are susceptible to mutation events in human tumors and which probably have an effect on structure and/or function of the p 21 proteins (see 3.3). It should be noted, however, that the mutation of ras oncogenes is rare. Mutations affecting the 12th amino acid of the c-ras-H-1 gene oncogene product occur infrequently in human cancer, since of 29 cancers, including 20 primary tumor tissues of bladder, colon, and lung, none revealed this alteration[134].

3.2.3 *Rearrangement*

The rearrangement of cellular oncogenes has been studied best in the case of the c-myc gene which is broken and translocated in Burkitt lymphoma. This will be extensively discussed below (4.3). The only oncogene for which rearrangement was demonstrated in a human tumor is c-abl, which is amplified and rearranged in the erythroid precursor cell line K 562[131].

The rearrangement of c-mos in mouse plasmacytomas is discussed in sect. 2.2.1. A rearrangement of c-myb oncogene in mouse plasmacytoid lymphosarcomas is a accompanied by abundant expression of c-myb mRNA of unusually large size[157].

Table 4. Detection of mutated ras genes in human tumors

Cell line/tissue	mutation (position/wild-type allele vs. tumor allele)	Ref.
related to c-ras-H-1		
T24/EJ bladder carcinoma cells[a]	12 (gly→val)	140, 142
lung carcinoma Hs 242 cells	61 (gly→leu)	143
melanoma	61 (gln→leu)	144
related to c-ras-N		
T-cell leukemias RPMI 8432, P-12, T-ALL-1	?[b]	135
melanoma cell lines (SK-MEL-119, SK-MEL-147)	?	136
AML (bone marrow)	?	137
promyelocytic leukemia HL-60	?	138
sarcoma cell lines RD, HT 1080	?	139
teratocarcinoma cell PA 1	?	17
neuroblastoma cells SK-N-SH	61 (gln→lys)	145
lung carcinoma cells SW-1271	61 (gln→arg)	141
related to c-ras-K-2		
lung adenocarcinoma A 549 cells	?	146
SCLL LX-1 cells	?	146
pancreatic carcinoma A 1165 cells	?	17
pancreatic carcinoma solid tumor 1189	?	147
rhabdomyosarcoma solid tumor 1085	?	147
lung carcinoma solid tumor 1085	?	147
colon carcinoma cells A 2233	?	147
gall bladder carcinoma cells A 1604	?	147
lung carcinoma A 427 cells	?	147
lung carcinoma PR 310 cells	not in first exon	148
lung carcinoma PR 371 cells	12 (cys→gly)	148
colon carcinoma cells SW 480	12 (gly→val)	149, 150
bladder carcinoma A 1698 cells	12 (gly→arg)	151
lung carcinoma cells A 2182	12 (gly→arg)	151
lung carcinoma solid tumor LC-10	12 (gly→arg)[c]	151
lung carcinoma Calu-1 cells	12 (gly→cys)[d]	149
ovarian carcinoma	?	152

[a] T 24 and EJ bladder carcinoma cell lines are probably identical[153)]
[b] not studied
[c] normal tissue lacked the mutated gene
[d] Calu-1 cell line is heterozygous at the c-ras-K-2 locus and carries one wild type allele (position 12: gly), and one mutated allele (pos. 12: cys)[148)]

3.3 Oncogene Products

There is only limited information about the products of cellular oncogenes. In most cases, the c-onc proteins have not been identified in normal cells. Even less is known about the subcellular localization of these products and their function. Table 5 summarizes the scarce data. It indicates the need for future studies along these lines. It is notable that the v-myb and v-myc gene products are nuclear proteins, however,

Table 5. Function and localization of viral and cellular oncogene products[a]

oncogene	localization and function of viral oncogene product		cellular oncogene product[b]		
			identified	localisation	function
abl			no	?	?
yes		tyrosine	no	?	?
src		kinases	yes	?	?
ros	plasma		no	?	?
fgr	membrane		no	?	?
fes/fps			no	?	?
erbB		structure of a tyrosine	yes	plasma membrane	EGF receptor
fms		kinase but no kinase	no	?	?
raf/mil(mht)	?	activity detected	no	?	?
mos	cytoplasm		no	?	?
sis	?	?	yes	secreted	PDGF
ras-H	plasma	bind GTP, auto-	yes	plasma membrane	bind GTP,
ras-K	membrane	phosphorylation	yes	(inner surface)	GTPase activity
fos	?	?	no	?	?
myb	nucleus	?	no	?	?
myc		binding of DNA	yes	?	?
erbA		?	no	?	?
ets		?	no	?	?
rel	cytoplasm	?	no	?	?
ski		?	no	?	?

[a] for detailed references see text

[b] data given for human cells; no other data can be added from other species except that a c-myb product has been identified in chicken cells

this has yet to be proven for the c-onc homologs (see sect. 3.1). The thus far best studied c-onc gene products are those of c-sis, c-erbB, and c-ras.

The v-sis is almost certainly derived from the gene (or genes) encoding a platelet-derived growth factor (PDGF)[155, 156], a notion strongly supported by the analysis of the human c-sis gene[157] and immunological studies[158]. PDGF is released from α-granules of platelets during blood clotting and it is the major polypeptide growth factor found in serum; fibroblasts and neuroglial cells are sensitive to the mitogenic action of PDGF, as they express PDGF receptors[159]. It is of interest that PDGF receptors exhibit tyrosine-specific kinase activity upon binding the growth factor[160].

Although the definite function of the c-ras gene products is not known, a number of data have been accumulated in order to elucidate the biochemical properties of the wild-type and mutant proteins. The ras proteins are all of a molecular weight of about 21000 D, hence, p^{21} proteins. The half-life of the c-ras-H-1 gene product was found to be 20 h in both the normal and mutant forms[161]. $p^{21\text{-H-1}}$ has a GTP binding capacity, which is not altered by the base-substitutions and amino acid alterations found in the tumor variants[162–164]. Similarly, their intracellular localization, as well as the degree of postranslational acylation is unchanged. The most notable structural difference is the observation of altered electrophoretic mobilities of some but not all mutated ras proteins[165], which concurs with the different three-dimensional structure of the mutated $p^{21\text{-H-1}}$ protein[166]; the mutant forms may be more conformationally restricted. The biochemical property of $p^{21\text{-ras}}$, which is clearly affected in the tumor forms is the GTPase activity, which is reduced in the mutated $p^{21\text{-H-1}}$ protein[162, 163]. The p^{21} proteins are located at the inner surface of the plasma membrane, with the membrane association and lipid binding mediated by the carboxyl terminus[167].

Utilizing antisera to a synthetic peptide corresponding to the carboxyl terminus of the human c-myc protein, a 48,000 D protein was precipitated from normal and tumor cells; this was taken to mean that a major change in the gene product may be no relevant factor in determining transformation[168].

The structural gene for c-erb-B is believed to code for the epidermal growth factor receptor[169]. This conclusion was derived from comparison of protein sequences and is further strengthened by the finding that the c-erb-B locus maps to 7p13→q22[88]; the EGF receptor gene was located independently to 7p22→qter[170, 171]. The binding of EGF to the EGF receptor triggers DNA synthesis with a latency period of about 12–24 h; the intrinsic tyrosine-specific protein kinase is stimulated upon binding EGF (note the similarity to PDGF action)[172].

3.4 Cooperation of Oncogenes

The fact that oncogenes must cooperate to fully transform normal into malignant cells was already suggested by transfection experiments. Cooper and Neiman[173] noted that DNA from tumors induced by avian lymphoid leukosis virus (LLV) contained transforming genes that can be detected by transfection of NIH 3T3 cells. Since these transforming sequences were not linked to LLV DNA sequences, it was concluded that oncogenesis by LLVs may result from indirect activation of cellular transforming genes. In a subsequent study[174)], these authors showed that LLV integrated near c-myc and that the transforming sequence was different from this locus. In Abelson leukemia virus (AbMuLV) induced lymphoid neoplasms, cellular transforming genes are detectable by transfection which are distinct from AbMuLV sequences[175)], suggesting that v-abl expression may mediate early events in the neoplastic process and that the generation of dominant acting transforming genes is secondary to this initial lesion. An analogous observation was made by Murray et al.[176)] in the human promyelocytic HL-60 cell line, where they detected the coexistence of mutated c-ras-N genes and altered versions of the c-myc gene. In Burkitt lymphoma cell lines, the c-myc is altered (see 4.3) and a dominant transforming gene (Blym-1) can be detected in the NIH 3T3 assay[15)]. Another hint for the importance of oncogene cooperation is the existence of retroviruses containing more than one oncogene (see 2.2.2). Lane et al.[177)] have studied the effects of transfection with a mutated c-ras-H-1 gene of mouse embryo fibroblasts. No conversion to tumor cells occurred; however, when the fibroblasts were established and immortalized before transfection, tumorigenic conversion did occur. Mortal embryo fibroblasts become tumorigenic if a second oncogene, such as viral or activated cellular myc gene (under the control of the SV 40 early promotor) or the gene for the polyoma large-T antigen is introduced together with the activated c-ras-H-1. That fibroblast immortality is a prerequisite for transformation by the mutated c-ras-H-1 gene has also been shown for primary dipolid hamster dermal fibroblasts, induced to grow indefinitely in culture by chemical carcinogens[178)]. In independent experiments, Ruley[179)] showed that the polyoma virus middle-T and the mutated c-ras-H-1 genes are individually unable to transform primary baby rat kidney cells. The adenovirus early region 1 A provided functions required by these genes to transform primary cells following DNA transfection. The same is true for Herpes simplex type 2. It was demonstrated that immortalization and tumorigenic transformation of early passage, non-established Syrian hamster embryo cells are genetically separable, and that cooperation between these two segments of HSV-2 genome is required for complete transformation[180)]. Polyoma middle-T antigen requires the cooperation from either small-T, SV 40 large-T, or Ad2 E1A genes, respectively to induce tumors in newborn rats[182)].

These data not only allow to definitely separate establishment and transforming function at the molecular level, they also suggest the existence of complementation groups for these two activities, the cooperation of which results in the full tumorigenic conversion of normal cells (Table 6).

In the following, I will operationally dissect some of the steps leading to the immortalization of cells, and will then discuss a possible model for the action of oncogenes in the transformation step of tumorigenesis.

It can be suggested that constitutive expression of proliferation genes leads to immortality. At present, it is not known what governs the conversion from regulated to constitutive expression of these genes. There might be irreversible effects (such as the activation events damaging the c-myc gene in Burkitt lymphoma cell lines, see 4.3), as well as reversible events. The latter have been evidenced in a number of experimental systems[188–190)]. It is clear, however, that the induction of proliferation genes is only the end result of a complex interaction of many regulatory circuits. In growth factor induced cell proliferation (with DNA synthesis occuring hours after

Table 6. Genetic complementation groups and genetically separable events in the multi-step genesis of human neoplasia

Step	Alteration of life-span	Malignant transformation	Progression
Genes involved	polyoma large-T[177] adenovirus-2 E 1 A[179, 182) a] herpes simplex-2 subgenomic fragment[180] v-myc[177) a] activated c-myc[177] polyoma small-T[182] SV 40 large-T[182] cellular "establishment and immortalization gene(s)"[d]	activated c-ras genes[177–179) b] v-ras-H[177)d] polyoma middle-T[179) c,b] herpes simplex-2 subgenomic fragment[180]	?

[a] Gene products bind nuclear structures[183–186]. A structural homology of the protein products of the myb and myc oncogenes and adenovirus-12 E1A genes has been noted[187]. This supports the previous argument that c-myc and c-myb are involved in cell.proliferation control (see 3.1)
[b] Gene products located at the plasma membrane (see 3.3)
[c] Polyoma virus middle-T gene is a dominant transforming oncogene in the NIH 3T3 transfection assay when linked to a murine retrovirus[181]
[d] These gene(s) have not yet been defined

the addition of growth stimulus to the cells), the initial event is the interaction of growth factor with growth factor receptors. This activates the protein tyrosine kinase activity of these receptors and leads to specific phosphorylations as key elements in the mediation of growth factor action[191]. What follows is a somewhat complex answer of the intracellular message system. It should be recalled here that there are oncogenes the products of which have functions in the above described path; there is a growth factor (sis), growth factor receptor (erbB) and tyrosine kinase (src-family) function represented in cellular oncogenes (see 3.3). Some years ago, it has been shown that a close association exists between rapid inositol lipid turnover and cell proliferation[192]. The breakdown of phosphatidylinositol 4,5-biphosphate results in the generation of 1,2-diacylglycerol and inositoltriphosphate. These two messenger molecules activate, respectively, protein kinase C (which is also the target of tumor promotors, Ref. 193) and the mobilization of Ca^{2+} ions from an intracellular store[194]. The observation that certain oncogene products can enhance the synthesis of phosphatidylinositol 4,5-biphosphate[195] is particularly intriguing. In this view, it may be anticipated that oncogene products that control, have or may have kinase activity will serve to control phosphatidylinositol 4,5-biphosphate supply, whereas other oncogene products might be involved in the breakdown of this molecule (receptors, phosphatidylinositol 4,5-biphosphate diesterase, or protein kinase C). In either case, this leads to a strong proliferative signal.

Observations that link the mutated ras genes to the control of cell transformation are as follows: In the case of the adenylate cyclase system, binding of the relevant hormone to the cellular receptor promotes the inactive form of the G-protein to be converted into the GTP state which in turn activates the cyclase. cAMP then stimulates a specific cAMP dependent protein kinase which performs the specific regulatory phosphorylations that determine the response of the cell. The G-protein signal is cancelled by slow hydrolysis of GTP to GDP, mediated by the GTPase activity of G-protein[196]. If ras gene products act as G-protein like molecules, then the loss of GTPase activity in the mutated forms (see 3.3) leads to a permanent activation of the adenylcyclase, with elevated levels of phosphorylation reactions.

To conclude, it is now quite clear that the multi-step origin of neoplasia, involves at least two different genetically determined processes, immortalization and transformation. The fact that cellular oncogenes and their products are involved at all levels of proliferation control (growth factor, growth factor receptor, intracellular message system) suggests that the exact nature of cell proliferation might be amenable to study with the genetic probes available to date.

4 Genetics of Specific Diseases

4.1 T- and B-Cell Leukemia/Lymphoma

Some leukemias and lymphomas carry a specific t (11 q 13;14 q 32) translocation; the DNA from a CLL patient with this translocation was used to demonstrate that the breakpoint on chromosome 14 occurs in the region carrying the immunoglobulin heavy chain locus[201]. A DNA probe that is specific for chr. 11 and that maps immediately 5′ to the breakpoint on the $14q^+$ chromosome was isolated and used to detect rearrangement of its homologous sequence in a case of diffuse large cell lymphoma with a t(11;14) translocation[202]. The authors concluded that the gene located on 11 q 13 and represented in their probe is involved in the malignant transformation of human B cells carrying the t(11;14) translocation; the gene, termed bcl-1, may be activated upon translocation into the heavy chain gene locus.

The fact that leukemias and lymphomas represent clonal expansions of malignant cells can be exploited in order to characterize such neoplasias on the basis of genetic markers for lymphocyte differentiation. The glycoproteins produced by B-cells, immunoglobulins, recognize and bind free antigens and are responsible for humoral immunity, while the T-cell molecules, T-cell receptors, recognize cell-bound antigens in the specific molecular context of self major histocompatibility complex products and are responsible for cellular immunity. In order to function in this way, B-cells make use of immunoglobulin genes, which have to be rearranged during B-cell differentiation in order to provide functional heavy and light chain genes. This rearrangement has been observed to take place in an ordered fashion, in which immunoglobulin heavy (μ) gene rearrangements generally preceed that of the light chain genes. Rearrangement of λ light chain genes follows that of κ[203], although exceptions to this model may exist[204]. In application of these findings, a number of clinical problems were studied. Korsmeyer et al.[203] showed that even though most leukemic non-T/non-B cells had not yet acquired the capacity to synthesize detectable cytoplasmic immunoglobulin (the earliest phenotypic characteristic of B-cells), they have undergone immunoglobulin gene rearrangements that may already commit them to B-cell development. The precursor cells that monoclonally expand to produce a lymphoid crisis in CML patients represent discrete steps in early B-cell maturation rather than the multipotential stem cell precursor[205, 206]. Immunoglobulin rearrangements have also been used as a diagnostic criterion in B-cell lymphoma and lymphoproliferative lesions[207, 208] and leukemias[209]. Thus, Ig-gene probes help to identify the leukemia/lymphoma cells of B-cell origin and allow classification of the degree of maturation.

The functional differentiation of T-cells seems to involve an analogous process of rearranging T-cell receptor genes[210], although this process is less well understood at present. It can, however, be anticipated that a T-cell receptor DNA probe will allow positive identification of malignant cells of the T-cell lineage, thus complementing the phenotypic characterization of T-cell malignancies in current use. The only cell which would escape such characterization with both Ig-gene and T-cell receptor gene probes is the lymphoid precursor cells which has not yet performed any rearrangement. The identification of such malignancies by genetic exclusion diagnosis in the above described manner should be of considerable clinical and biological interest.

4.2 *Chronic Myelogenous Leukemia*

In most cases of CML, a marker chromosome (Ph′) is present in the various cell lines involved in the disease (granulocytic, erythrocytic, megakaryocytic, and lymphocytic), suggesting that all malignant cells in this disease are derived from a single multi-potent stem cell clone. The Ph′ chromosome arises from a balanced translocation between the long arms of chromosomes 9 and 22, with break points in bands q 34 and q 11, respectively[73]. de Klein et al.[74] and Rabbitts et al.[75] showed that c-abl sequences are translocated from chromosome 9 to chromosome 22 q. Subsequent studies[211] indicated that the c-abl gene is not translocated in Ph′-negative CML, supporting clinical evidence that CML without Ph′ represents a distinct entity[212]. The c-sis gene, localized at 22 q 12.3 – q 31.1 (far away from the breakpoint region 22 q 11) segregates with the translocated part of chromosome 22 to different chromosomes in Ph′-positive patients and remains on chromosome 22 in Ph′-negative cases[76]. It is notable that the breakpoint on chromosome 9 is variable in different Ph′-positive CML patients and may be found within a relatively large but limited region on chromosome 9[213, 214] associated with the c-abl locus. It is not certain, whether these breakpoints lie within the c-abl transcriptional unit and/or regulatory domain, since the human c-abl has not been defined in sufficient detail as yet. Data from many laboratories including our own have failed to demonstrate a rearrangement of c-abl sequences complementary to v-abl gene probes in CML DNA[73, 74, 214] (to be published). In contrast to the situation on chromosome 9, the breakpoints on chromosome 22 are clustered in a very limited region, termed bcr, of about 5.8 kb[214]. All Ph′ positive CML patients studied showed a chromosomal break within bcr, whereas Ph′-negative patients did not exhibit such a break. It further appeared that the rearrangements involving bcr are highly specific for the leukemic cells in Ph′-positive CML patients, and may thus be of diagnostic value, in particular, when no adequate metaphase chromosome spread can be obtained.

The molecular and cytogenetic studies of CML now firmly establish that in Ph′ positive cases the c-abl and bcr, but not c-sis are the critical sites involved in the genesis of this disease; however, the functional consequences of rearrangements at these sites remain to be elucidated.

4.3 Burkitt Lymphoma

The Burkitt lymphoma (BL) is characterized by three different cytogenetic aberrations, all of which involve chromosome 8. These translocations t(8q24;14q32), t(8q24;22q11), t(2p11;8q24) are reciprocal exchanges of material with involvement of the c-myc locus on chromosome 8, and the immunoglobulin gene loci on chromosomes 2, 14, and 22, respectively. As shown below and summarized in Table 7, the c-myc locus is affected in BL in several ways, (i) the breakpoint can be on either site of c-myc, suggesting that it may or may not reside on chromosome 8 in tumor cells, and (ii) the c-myc locus can either become structurally changed by the translocation process or may be left apparently unchanged with the breakpoints several hundred base pairs to several thousand base pairs 5′ or 3′ of the c-myc locus.

The human c-myc locus has two exons with sequence homology to the viral v-myc oncogene. Recently, it was discovered that another first exon is also present. Figure 1 depicts a simplified scheme of the human c-myc locus, highlighting the features pertinent to the present discussion. It should be noted, however, that there is some controversy regarding the first exon and its potential coding capacity. Most workers who have sequenced the c-myc locus demonstrated multiple translational stop signals in all possible reading frames[215–218]. Gazin et al.[219] have found in their sequence an open reading frame allowing for a protein of about 20,000 D. The reason for this apparent controversy is unclear at present and must await further studies. One of the intriguing features of the c-myc gene is the presence of two promotors in the first exon region, and the fact that cryptic promotors are revealed in cases where the function of these upstream sites is destroyed[220].

Animal models of BL are the murine plasmacytomas; the mouse plasmacytomas characteristically have a translocation of the distal part of chromosome 15 to chromosome 12 (t(12;15)) or 6 (t(6;15)). Chromosome 12 harbors the immunoglobulin heavy chain gene locus, the immunoglobulin λ light chain gene resides on chromosome 6. The breakage of the c-myc gene occurs within the large 5′ non-coding exon or the first intron. This results, as in some cases of the common form of BL (most notably in the Manaca cell line), in the separation of the normal c-myc gene promotor sequences from the c-myc coding region[221, 222]. However, although truncated c-myc specific transcripts are generated, the amino acid compositions of the c-myc genes in normal and translocated forms are equivalent. Therefore, the gene product itself is qualitatively unaltered in mouse plasmacytomas[223]. This immediately implies that it is the quantitative difference in normal and plasmacytoma cells of the c-myc protein which determines the malignant behaviour of the tumor cells. In this respect, it is interesting to note several experiments concerned with the level of expression of the c-myc gene in BL cell lines. A small and variable increase in c-myc expression (2–5 fold) among undifferentiated B cell lymphomas of Burkitt and non-Burkitt type has been found[224]. This argues against increased transcription of this gene being the cause of BL, and suggests that altered regulation of expression of this gene might be a critical event in tumor formation. The normal

Table 7. Cytogenetic and molecular characteristics of Burkitt lymphoma cell lines

cell line	translocation	breakpoint relative to c-myc[a]	comments	Ref.
Raji	t(8;14)	5′	expression of c-myc elevated; c-myc broken in first exon; base-pair mutations in translocated c-myc coding exons; c-myc probably translocated first to cμ, then by pseudo-switch to cγ1; immunoglobulin enhancer (between cμ and J_H) not available for c-myc.	231, 233, 234
Manca	t(8;14)	5′	immunoglobulin enhancer element is translocated to within the first exon of c-myc and activates transcription from cryptic promotor sites in the first c-myc intron.	220
P3HR-1	t(8;14)	5′	c-myc locus structurally altered; c-myc expression elevated.	235
W1	t(8;14)	5′	c-myc locus structurally altered, c-myc expression *not* elevated.	236
IARC-BL22	t(8;14)	5′	c-myc locus structurally altered; coding exons unchanged; translocation breakpoint 1000 base pairs 5′ of c-myc promotors.	237, 238
IARC-BL31	t(8;14)	5′	c-myc locus structurally altered.	237
ST486	t(8;14)	5′	c-myc locus broken in first exon	239, 240
JI	t(2;8)	3′	c-myc locus in germ-line configuration, c-myc expression elevated.	241, 242
JBL2	t(2;8)	3′	breakpoint 20,000 base pairs downstream of the c-myc gene; abnormal expression of translocated allele (shift in promotor utilization, increase in transcript level); duplication of a 2,500 base pair segment of DNA containing the two c-myc promotors and exon 1.	243
IARC-BL37	t(8;22)	3′	c-myc locus is in germ-line configuration; translocation is reciprocal and relatively conservative; c-myc expression elevated (promotor shift).	244
IARC-BL12	t(8;22)	3′	c-myc locus in germ-line configuration; c-myc expression elevated.	245

[a] the human c-myc locus is orientated on chromosome 8 in the order centromer-5′-c-myc-3′-telomer[246)]; the point of internal reference is the ATG-codon of exon 2.

c-myc allele is repressed, the translocated c-myc allele active in B cells, suggesting that the altered c-myc allele escapes from the normal transcriptional control[225, 226)]. The fact that a differential B cell environment is required for the ability to express a productively rearranged immunoglobulin μ chain gene and the translocated c-myc gene suggests the involvement of specific trans-acting regulatory factors in transcriptional control of the c-myc gene[227, 228)]. It has already been mentioned that a c-myc gene product has been immunoprecipitated from Burkitt lymphoma cells, which has a normal molecular weight and electrophoretic mobility[168]. Thus, a major change in the c-myc protein is unlikely to be a relevant factor in transformation.

All these data suggest that the situation in mouse plasmacytomas is comparable to the situation in human BL. There are, however, certain dissimilarities which will be dealt with in the following.

In an analysis of t (8;14) forms of BL, Taub et al.[229] noted variable translocation breakpoints with respect to the c-myc gene. Whereas all rearrangements occured 5′ to the two coding exons, only 2 of 5 cell lines studied showed a disruption of the normal three exon structure of the human c-myc gene, which characteristically occurs in mouse plasmacytomas. In an attempt to elucidate the alteration in c-myc expression, the levels of two alternative c-myc transcripts (initiating at cAp sites 1 and 2, see Fig. 1) were compared. Normal B cells predominantly utilize promotor 2, whereas in BL cell lines, promotor 1 is much more frequently employed. This promotor shift is accompanied by transcriptional silencing of the normal, unaltered c-myc allele.

The loss or damage by rearrangement of the first exon is not a prerequisite for this effect. In cases, where the three exon structure of c-myc is left intact, the promotor mutations are exerted by base changes probably of the type that induce somatic mutations in the rearranged immunoglobulin genes of normal immunoglobulin producing cells[230]. Data suggesting that base substitutions are indeed a different mode of c-myc promotor mutation, secondary to the rearrangement, have also been reported by others[231].

We are thus left with the fact that in human Burkitt lymphoma, characteristic promotor lesions are introduced into the rearranged c-myc allele. This can be accomplished by direct breakage of the three exon structure of the gene as is the case in the mouse plasmacytomas, or by base-pair mutations induced by translocation of an immunoglobulin gene into the vicinity (either 5′ or 3′) of the c-myc locus. This promotor damage is accompanied by the loss of transcription from the normal c-myc allele, suggesting a trans-acting factor in regulation of c-myc expression. There are two different models which have been proposed to account for the observations in BL cells, one based on transcriptional, another based on translational control. Figure 2 schematically illustrates these models.

The transcriptional regulation model proposed by Leder and coworkers[232] indicates that the promotor region of c-myc is a site for interaction with a repressor-type molecule. This may either be the c-myc gene product itself (autorepression) or an unrelated protein. Recent data of Rabbitts et al.[231] suggest that the c-myc gene product is directly or indirectly involved in the regulation. They have found that unlike most other BL cells, the Raji cell expresses both the translocated and the normal c-myc allele. Raji cells possess a substantially altered translocated c-myc gene-coding segment suggesting that this c-myc gene product although it can mediate transformation is inefficient in repressing c-myc gene transcription.

Damage at the repressor binding site either by translocation from the c-myc coding portion or by base-pair mutations would prevent the repression to function properly, whereas the normal allele would be transcriptionally silent.

The translational regulation model proposed by Saito et al.[218] is based on the observation that the c-myc germ line mRNA can adopt a stem-loop type secondary structure, with the initiation codon AUG at the second exon within the loop. The c-myc mRNA transcribed from the translocated allele cannot form the stem-loop structure and is thus translated with high efficiency.

In conclusion, the reviewed findings strongly indicate that an altered control of expression of the c-myc gene, rather than quantitative changes are the crucial point in transformation of BL.

Although not directly related to BL, some recent data concerning the role of c-myc in proliferation control should be mentioned here in order to substantiate the above conclusion further. Campisi et al.[111)] showed that growth stimulation of quiescent Balb/c 3T3 mouse cells by serum elevates c-myc expression. In two chemically transformed derivatives, c-myc expression is constitutive. By contrast, the expression of c-ras-K is still cell-cycle dependent in the transformed cells. Although the c-myc gene is not rearranged, amplified, or overexpressed, it has lost its cell-cycle dependent regulation in chemically transformed cells.

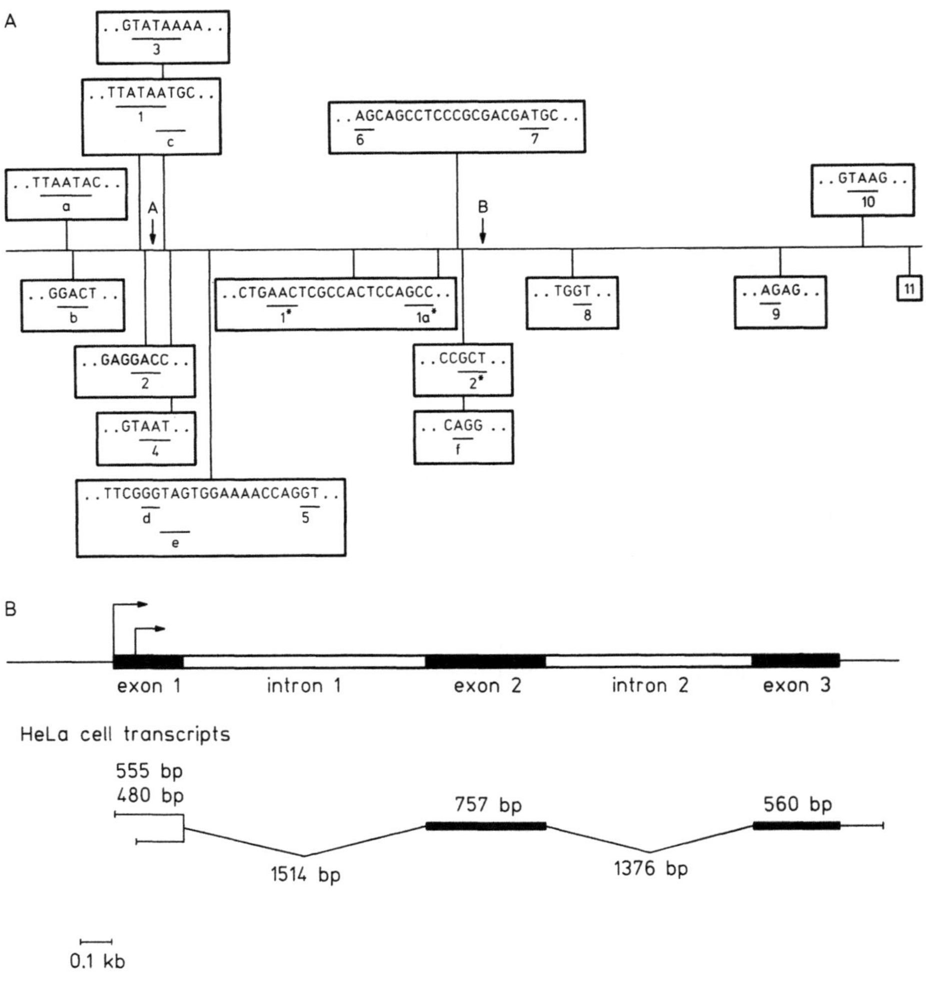

In other experiments, Kelly et al.[247] have specifically addressed the cell-cycle dependent induction of c-myc. Agents that initiate the first phase of a proliferative response in lymphocytes (lipopolysaccharide or Concanavalin A) and fibroblasts (platelet-derived growth factor) induce c-myc mRNA. Within one to three hours after the addition of these mitogens to the appropriate cells, c-myc mRNA increases 10–40 fold. The induction does not require protein synthesis. The fact that c-myc mRNA is even more elevated by the combination of mitogen and cycloheximide (a protein synthesis inhibitor) suggests the involvement of a labile protein which regulates c-myc mRNA in these cells.

It is of importance to recognize that mechanisms other than rearrangement and amplification can lead to an aberrant, uncontrolled expression of cellular oncogenes. It is not known what kind of mechanism (s) is responsible for that, however, it can be anticipated that alterations introduced by base-pair mutations or alterations in the pattern of DNA cytosine methylation are effective in this respect[248–250].

4.4 Retinoblastoma

Retinoblastoma most likely arises from embryonal cells and could result from as few as two specific genetic events, the first of which can be inherited. The gene for hereditary retinoblastoma (Rb+ for the wild type allele) was mapped to chromosome 13q14 by interference from a number of observations: (i) The constitutional deletion of the long arm of chromosome 13 always includes band q14[251, 252]. (ii) Deletions and rearrangements of chromosome 13, resulting in the loss of structural material from band q14, as well as monosomy of chromosome 13 has been noted in

◀ **Fig. 1. A** *Human c-myc locus.* The structure and the sequences of the human c-myc locus have been compiled from references given in the text. This scheme shows as numbers (1–11) the landmarks of the c-myc gene as suggested by the majority of investigators: 1, promotor 1; 2, cAp site 1; 3, promotor 2; 4 cAp site 2; 5, first splice donor site; 6, first splice acceptor site; 7, initiation codon; 8, second splice donor site; 9, second splice acceptor site; 10, stop codon; 11, polyadenylation site. The numbers with asteriks (1*, 1a*, 2*) denote the start sites of transcription in Manca cells. The letters A and B refer to an internal sequence complementarity seen in exon 1 and exon 2[218]. The letters a–f mark features observed by Gazin et al.[219], who obtained a slightly different sequence than other authors. This allowed an open reading frame in exon 1 and the appropriate signal sites are as follows: a, putative TATA box; b, putative cAp site; c, putative initiation codon; d, putative donor splice site; e, putative stop codon; f, putative alternative splice acceptor site.
B *Transcripts from the human c-myc locus.* HeLa cell transcripts contain 555 bp or 480 bp of the first exon, depending on the promotor used for transcription initiation, about 1514 bp from the first intron, 757 bp from the second exon, 1376 bp from the second intron, 560 bp from the third exon, and about 300 bp 3′ sequences ending with the poly(A) tail. Manca cell transcripts initiate at sites 1*, 1a*, or 2* and run through the first splice acceptor site of the germ-line gene as above. The putative transcripts proposed by Gazin et al.[219] would initiate at b, and would run through the complete c-myc locus. Note, that in their sequence, an initiation codon (ATG) is present just at the site where promotor 1 is located by other investigators. As explained in the text, the discrepancy in reported sequence data is not yet resolved. Since the sequence of Gazin et al.[219] also contains additional insertions in the sequence downstream the promotor 1 site, it would allow an open reading frame of about 450 bp, with coding capacity of about 20,000 D. The presence of the stop codon at e would impair the production of the protein encoded in exons 2 and 3, unless an aberrant splicing would occur between d and f, possibly allowing the synthesis of a 68,000 D protein

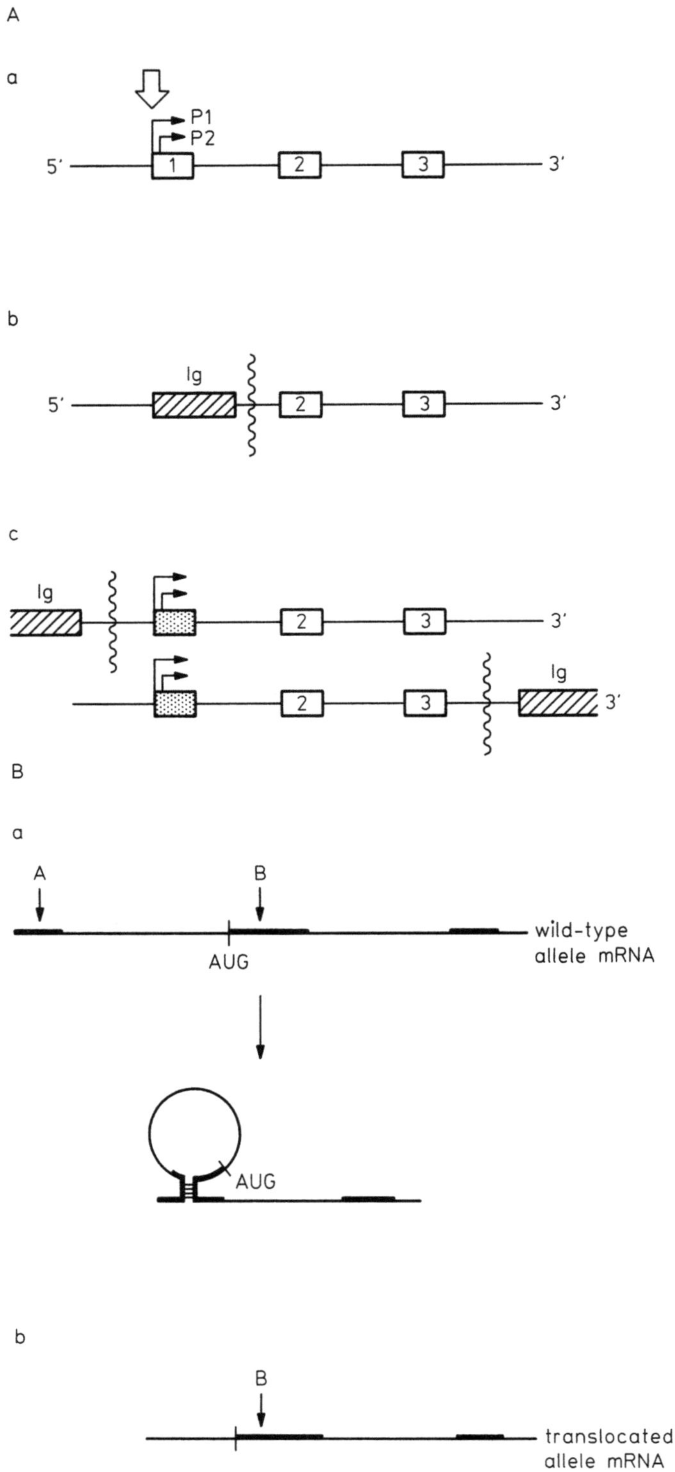
A
a
P1
P2
5'
1
2
3
3'
b
Ig
5'
2
3
3'
c
Ig
2
3
3'
2
3
Ig
3'
B
a
A
B
AUG
wild-type
allele mRNA
AUG
b
B
translocated
allele mRNA

Table 8. Characterization of retinoblastoma genotypes at the Rb-1 locus in 13q14

sporadic form[a]		hereditary form[b]		chromosome 13 deletion form	
constitutional	tumor	constitutional	tumor	constitutional	tumor
Rb+/Rb+	rb–/rb–[d] or rb–/–[e]	Rb+/rb–[f]	rb–/rb– or rb–/–	Rb+/–	rb–/– or –/–

[a] no positive family history, disease usually unilateral
[b] positive family history, bilateral, multifocal disease
[c] characterized by loss of chromosomal region 13q14
[d] homozygosity of the mutated allele
[e] hemizygosity of the mutated allele
[f] carrier of the retinoblastoma mutation, the risk of developing retinoblastoma is increased by more than 100,000 times[262)]

tumor cells of karyotypically normal individuals[253)], with similar abnormalities in cells from patients with both the sporadic and heritable forms of the tumor. (iii) The locus for esterase D was assigned by deletion mapping to the chromosomal region 13q14 and found to be tightly linked to the gene of hereditary retinoblastoma[254)] suggesting a common underlying mechanism in the pathogenesis of the chromosomal deletion and hereditary forms of retinoblastoma. Thus, all evidence suggests that band q14 on chromosome 13 contains a recessive gene of prime importance in the development of retinoblastoma. A substantial evidence now suggests that tumorigenesis may result from the development of homozygosity for the mutant allele (rb-) at the Rb-1 locus, suggesting a two-step model of tumorigenesis consisting of a combination of an inherited mutation and a somatic mutation, an idea put forward originally by deMars[255)] and applied for the case of retinoblastoma by Knudson[251)]. Table 8 illustrates this model. From studies by several groups, combin-

◀ **Fig. 2. A** Regulation of c-myc transcription in Burkitt lymphoma cells *(a)* The c-myc gene is regulated by the action of a repressor (thick arrow), which, as discussed in the text, directly or indirectly, depends on the presence of the normal c-myc product. The target site for the repressor is the dual promotor region. This type of autorepression is altered if the target site is damaged. Note that the c-myc locus is not drawn to scale. *(b)* Separation of promotor region from the coding exons 2 and 3 by rearrangement results in the loss of the triple exon structure of c-myc. The target site for the repressor is lost. This activates cryptic promotors in the c-myc gene region (first intron), and releases the rearranged c-myc allele from trans control, whereas the wild-type allele is repressed. *(c)* Loss of promotor function by somatic mutation is induced by rearrangement of the immunoglobulin (Ig) gene loci 5′ or 3′ to the intact c-myc gene region. While the triple exon structure is left intact, the promotor region suffers from base-pair mutations (indicated by the stippled first exon), which result also in the loss of the interaction site with the repressor. Mutated c-myc allele is released from trans control, wild-type allele is repressed. **B** Regulation of c-myc translation in Burkitt lymphoma cells *(a)* The wild-type allele gives rise to a large transcript, which contains the complementary sequences A and B. Thus, the germ-line c-myc mRNA can adopt a stem-loop type secondary structure, with the initiation codon at the first coding exon within the loop. *(b)* The c-myc RNA from the translocated allele cannot form the stem-loop structure and is thus believed to be translated with high efficiency. It should be noted, however, that the translational regulation model of c-myc expression can only apply to cases where the triple exon structure is damaged. In all other cases, where the promotor sites still function in the initiation of transcription but are defective in binding the repressor, such a model could not apply

ing cytogenetic[256, 257] and restriction enzyme[258] analyses, it now seems clear that the development of homo- or hemizygosity, respectively for the mutant allele at the Rb-1 locus at 13q14 includes mitotic non-disjunction with loss of the wild-type chromosome 13, or mitotic recombination events. There are, however, cases, for which these two mechanisms do not apply. Thus, though not yet formally demonstrated, a point mutation or a small deletion at the homologous Rb-1 locus, or a regionalized recombinational event such as gene conversion, could all reveal the recessive mutant allele at the cellular level. Utilizing this information, makes the hereditary form of retinoblastoma accessible to genetic counseling, with 13q14-specific markers (e.g. esterase D[254], or hybridization probes[258]) available. If the tumor has lost the alleles present on one chromosome 13 homologue, one can assume that the chromosome 13 remaining in the tumor is the one carrying the mutant allele. In the patient's family, the allele predisposing for retinoblastoma is thus identified[259].

Murphee and Benedict[260] have argued that as a result of the loss of the "suppressor function" of the Rb- wild type allele, the regulation of "expressor gene(s)" might be impaired. They suspect the presence of such gene(s) on chromosome 6p, since an i(6p) is present in several retinoblastomas from patients with both the hereditary and non-hereditary disease as a fairly specific marker chromosome.

The case of retinoblastoma thus illustrates the concept of Comings[261] who suggested that inactivation of suppressor genes may play a role in human tumorigenesis.

The molecular cloning of the Rb-1 locus will provide insights not only into the nature of the putative suppressor gene but also into the precise mechanism(s) of inactivation events at this site. The DNA of patients carrying both the normal and mutated alleles of the Rb-1 locus will be the source for this task.

Wilms' tumor is also a childhood tumor which has hereditary, non-hereditary, and chromosomal deletion forms, with the affected gene at 11p13[263], found to be frequently deleted in tumor cells. It has been shown that the c-ras-H-1 locus, as well as the insulin and β-globin gene loci map outside the deletion associated with these tumors[264, 265].

5 Addendum*

An interesting addition to the *assay systems for transforming activities of tumor DNAs* are JB6 mouse epidermal cells. These cell lines include preneoplastic clones that are sensitive to promotion of transformation, variants that are resistant to promotion of transformation, and tumorigenic lines produced by tumor promotor exposure of promotion-sensitive lines. By transfecting DNAs from such sources into JB6 clones, it was shown that the genetic activities specifying the sensitivity to

* Since the preparation and submission of this mansucript, a number of significant contributions pertinent to the present discussion have been made. The purpose of this addendum is to briefly discuss these findings in the light of the previous observations

promotion of transformation by tumor promotors and the transforming activity present in tumor DNAs are different[266].

The characterization of integration sites of *non-acutely transforming retroviruses* continued to reveal new information on these putative cellular oncogene loci in the mouse genome. In early T-cell lymphomas, proviral integration of MuLV can be observed simultaneously near c-myc and pim-1 in the same cell clone; the significance of this observation has yet to be elucidated[267]. MuLV integrates into Mlvi-1 and Mlvi-2, two DNA regions involved in the induction of mouse thymic lymphomas; these loci were both mapped to chromosome 15, and found to be distinct both from each other and from c-myc and c-sis, also found on this chromosome in mouse[268]. The putative oncogene int-1, activated by integration of MMTV has been molecularly cloned; this study revealed that the protein-coding domain is always left intact by the integration process[269]. This provides further evidence that the int-1 specified protein contributes an essential step in mammary tumorigenesis.

During the *characterization of virally transduced oncogenes,* it was found that erbA is related to carbonic anhydrases, which participate in the transport of carbon dioxide by erythrocytes[270]. AEV can specifically block the maturation of erythroid cells at an early stage of differentiation, however, this requires the synergistic action of erbB (a src-family related oncogene) and erbA (probably an activated specific cell-derived gene), whereas erythroblasts transformed by src or fps alone require complex growth conditions and partially differentiate into mature erythrocytes in vitro[271].

A *new family of oncogenes* was described in experiments using rat neuro/glioblastoma cellular DNA in the NIH 3T3 transfection assay. The new oncogene encodes a 185 kD tumor antigen (p185) which is serologically related to but distinct from the epidermal growth factor receptor[272, 273].

Additional studies of constitutive *fragile sites* still do not unequivocally clarify the critical question whether fragile sites are more than just predisposing factors for chromosomal rearrangements, or may constitute sites of genes critical to the genesis of human cancer[274, 275]. The *chromosomal localization of genes relevant to tumorigenesis* has been further advanced. The human c-fos gene was mapped within chromosomal region 14q21–q31[276]. It was shown that only one of the two src-related loci, namely c-src-1, gives rise to a detectable mRNA[277]. The chromosomal localization of the human T-cell antigen/MHC-receptor (T_i) genes $T_i\alpha$ and $T_i\beta$ have been mapped to chr. 14, and chr. 7q22-qter, respectively[278–279a]. This is of note, since one might expect the involvement of T_i genes in cases of T-cell malignancies, in analogy to the B-cell derived lymphomas with their characteristic translocations involving the Ig gene loci.

Studies concerned with the *expression of oncogenes* have added new information to the interrelationship between cell proliferation and differentiation. By analysing the spatial and temporal expression of c-myc in developing human placenta, a strong correlation between c-myc transcript abundance and cytotrophoblast proliferation was found[280]. A close link between reduction of c-myc expression and G_0/G_1 arrest has been established in experiments using interferon β-sensitive and -resistant subclones of human Daudi cells[281].

Thus, all cells which become arrested in the G_0/G_1 phase of the cell cycle as part of their terminal differentiation might be characterized by a decline in c-myc expression. The differentiation of HL-60 cells into macrophage-like cells by TPA (12-O-tetradecanoylphorbol-13-acetate) is accompanied by the induction of both c-fos mRNA and protein within 15 min after treatment, whereas such events could not be observed in the monocytic and granulocytic differentiation processes of HL-60 cells[282)], supporting the notion that c-fos may serve macrophage-specific functions. Further, c-fos expression is low in osteogenic cells, but high in bone marrow and macrophages; in the latter, it is confined to mature cells[283)]. In mouse fibroblasts, serum or purified growth factors induce a rapid (within 10 min) but transient increase in mRNAs specific for c-fos and a c-fos related gene, followed by the synthesis of nuclear c-fos proteins[284–287)]. This induction of c-fos precedes that of c-myc[288)]. In macrophage proliferation, expression of c-fos is also inducible by the macrophage-specific growth factor CSF-1; however, the kinetics are entirely different from those in growth-factor stimulated fibroblasts[282)]. Although c-fos can be induced rapidly in fibroblasts and HL-60 cells by TPA, the end result of this induction is different, yielding to proliferation in the first and to differentiation in the latter case. It is conceivable that the c-fos product functions as an intracellular mediator of TPA action at the cellular membrane, the ultimate outcome of this triggering process depending on a specific cellular commitment. The N-myc gene, frequently found to be amplified in neuroblastoma cells (see 3.2.1), is expressed more actively per genomic copy in such cells than in non-neuroblastoma cells and tumors[289)]. Its expression decreased during retinoic-acid induced morphological differentiation of human neuroblastoma cells prior to the appearance of morphological changes[290)]. This suggests that N-myc may be a cell-type specific proliferation gene for neurogenic tissues.

The *amplification of oncogenes*, namely c-K-ras-2 and c-myc has been found simultaneously in a human lung giant cell carcinoma LU-65, which is maintained in nude mice. Both genes were amplified 8–10-fold, which was accompanied by a point mutation in c-K-ras-2[291)]. Three of sixteen samples of human gastric adenocarcinoma carry amplified c-myc genes with a concomitant increase in c-myc mRNA. Karyotype analyses revealed the presence of double-minute chromosomes[292)].

Some progress has also been made in the *characterization of oncogene products*. The c-myc gene product has been identified as a nuclear protein of molecular weight 65 000 D[293)]. In subsequent studies, the same group of investigators used antisera prepared against synthetic peptides representing different portions of the c-myc protein, to immunoprecipitate c-myc related proteins from mouse, rat, hamster, frog, and monkey cells. This suggests an evolutionary conservation of antigenic structure, and possibly function, of the c-myc protein among vertebrates[294)]. The human c-myc gene product is a DNA-binding protein[293)], just as a bacterially made homolog[295)]. More detailed studies with regard to the human c-myc protein have been carried out by others. Ramsay et al. identified two phosphoproteins, $pp^{62c\text{-}myc}$, the major product, and $pp^{66c\text{-}myc}$, produced in smaller quantities[296)]. The half-live of these two proteins is about 30 min[296)].

Hann and Eisenman found two nuclear phosphoproteins of molecular weights 64 and 67 kD as human c-myc products, which were shown to be highly related by proteolytic mapping [297)]. Interestingly, an alteration in the expression of $pp^{64\,c\text{-}myc}$ relative to $pp^{67\,c\text{-}myc}$ was observed, paralleled by differences in promotor utilization of promotors P1 and P2 of the c-myc locus (see Fig. 1.A). This was achieved by comparing normal cells and cells carrying structurally altered c-myc alleles.

A large number of studies is still concerned with the biological functions of ras oncogene products. The decreased GTPase-activity of mutated p^{21} ras proteins has been substantiated by three other groups[298–300]. The requirement of Ha-ras p^{21} protein for transformation of NIH 3T3 cells was directly demonstrated by microinjection of the bacterially made oncogene product into recipient cells[301]. In addition, it was observed in this study that the injection of Ha-ras p^{21} stimulated quiescent fibroblasts to enter the S-phase of the cell cycle, an observation independently made by others[302]. A second important evidence for the transforming potential of Ha-ras p^{21} proteins is an experiment involving the injection of anti-ras antibodies into fibroblasts. When antibodies specific for amino acid 12 were injected, a transient reversion of oncogene-induced cell transformation was observed[303]. Injection of monoclonal antibodies against ras proteins showed that these proteins are required for the initiation of the S-phase in fibroblasts[304]. The transforming activity of ras-proteins is mostly localized to the N-terminus, whereas the C-terminus may function in the recognition of specific stimuli, differing for each of the ras protein types[305, 306]. Alterations in the electrophoretic mobilities of position-12- and position-61-altered p^{21} molecules are different from each other and wild-type products, allowing a rapid immunological assay for ras mutations[307]. Monoclonal antibodies of predetermined specificity have also been used to detect mutated p^{21} ras proteins in immunochemical studies[308]. The tumor, but not the normal tissue of two patients with lung and mammary cancer contained mutated forms of the c-Ki-ras-2 and c-Ha-ras-1 alleles, respectively; this suggests that the activation of ras oncogenes may concur with the development of human neoplasia[309, 310]. Mutations in ras genes, however, may also be acquired after establishment of human tumor cells in culture, as shown for the human PA-1 teratocarcinoma cell line[311], a finding reminiscent of the oncogene activation in ras loci during the progression of a mouse lymphoma[312].

The evolutionary relationship of the epidermal growth factor receptor gene has been worked out in some detail. These studies show that (i) the second extracellular domain has a homology to α_1-acid glyco-protein (known to be distantly related to immunoglobulins), (ii) the transmembrane domain is similar to that of the HLA class II β-chain, and (ii) antibodies to two defined regions of the transforming regions of the transforming protein pp^{60src} interact specifically with the epidermal growth factor receptor kinase system located in the cytoplasmic domains[313, 314]. The gene for the EGF receptor is amplified, expressed at elevated levels, and possibly rearranged in primary human brain tumor cells of glial origin (A-431 cells)[315].

Studies on the *cooperation of oncogenes* and their classification into genetic complementation groups (Table 6) have revealed another immortalizing gene, p53. The cellular tumor antigen p53, is formed at elevated levels in a wide variety of transformed cells[316], although its relationship to malignant transformation is still not fully understood. Certain similarities with the product of c-myc exist, since both are regulated with the cell cycle[111, 247, 317], and the amino acid sequences predicted for the two proteins show weak similarities in both the overall molecular organization and the positioning of charged residues within distinct domains of the molecules[318]. Indeed, it was shown that the protein encoded by the p53 gene can complement activated ras genes in the transformation of primary rodent cells[318, 319]. Its classification as an immortalization gene was directly demonstrated with xiphisternum chondrocytes (WAXI cells) by Jenkins et al.[320]. The Ad2 E1A protein is a nuclear protein[321], and able to stimulate both viral and cellular gene expression[322].

The transmission of the polyoma middle T gene as the dominant acting oncogene of an in vitro recombined murine retrovirus definitely proves its assignment to the group of transforming genes[323]. The homologies between signal transducing G proteins and ras gene products suggested in light of their GTP binding and hydrolyzing activities has been substantiated by a comparision of primary sequences of the α subunit of a bovine brain G protein and the α subunit of rod outer-segment transducing with ras gene products; the NH_2-terminal peptide sequences of these proteins are highly homologous[324].

Keath et al.[325] constructed fibroblast lines expressing activated c-myc oncogenes and have shown that these cells are tumorigenic in nude mice and syngeneic animals; their data suggest that constitutive c-myc expression leads to tumorigenicity in immortalized cell lines. In transgenic mice that carry an otherwise normal mouse c-myc gene, in which increasingly larger portions of the myc promotor region have been replaced by a hormonally inducible MMTV promotor, spontaneously developing mammary adenocarcinomas were observed[326]. Not only the female founders of the transgenic strains but also F1 female progeny developed these tumors during their early pregnancies. These studies provide evidence that a constitutionally deregulated myc gene appears to act as a heritable, predisposing factors, favoring the accelerated development of a tissue-specific adenocarcinoma. Molecular studies with *T- and B-cell leukemia and lymphoma* revealed some additional clues regarding the mechanisms underlying the genesis of these malignancies. Studies with an acute B-cell leukemia cell line with 14;18 and 8;14 chromosome translocations allowed the identification of a second putative B-cell lymphoma oncogene locus (termed bcl-2) or chr. 18q21[327, 328]. Data suggesting that the switching process proceeds in the general order consistent with the proposed order of the heavy chain gene classes on chr. 14[329] (namely $5'$-$(V_H)_n$-D-J_H-C_μ-C_δ-$C_{\gamma3}$-$C_{\gamma1}$-Ψ_ε-$C_{\alpha1}$-Ψ_γ-$C_{\gamma2}$-$C_{\gamma4}$-C_ε-$C_{\alpha2}$-$3'$, where Ψ_ε and Ψ_γ are pseudogenes[330]), may allow a rather subtle characterization of monoclonal expansions of B-cells. By use of a cDNA probe for the $T_i\beta$ subunit of the T-cell antigen/MHC receptor, it was found that $T_i\beta$ rearrangement is evident in stage II (T11 + T6 + T3 −) and stage III (T11 + T6 − T3 +), but not stage I (T11 + T6 − T3 −) thymocytes; the expression of $T_i\alpha$ and $T_i\beta$ molecules (the heterodimer representing the T-cell receptor) was restricted to stage III thymocytes[331]. Thus, human T-lineage ontogeny is, analogous to the B-lineage, characterized by an orderly series of differentiation steps. This can be exploited to molecularly characterize human T-cell leukemias[331–333].

Chronic myelogenius leukemia has now been additionally characterized by the finding of aberrant transcripts from the c-abl locus situated on chr. 9q34.1. The finding of an 8kb abl transcript, as opposed to the normal 6 and 7kb transcripts, was associated with the presence of the Ph′ chromosome, i.e. the characteristic 9;22 translocation[334–335a]. The data suggested that the aberrant transcript contained extra information from the region 5′ of the translocated oncogene. The transcript might be translated into an altered protein, perhaps modified at the NH_2-terminus region. This region has been previously shown to be critical for the transforming activity of the protein encoded by v-abl[336].

The rearranged c-myc gene from the *Burkitt's lymphoma* cell line CA-46 is characterized by the fact that the non-coding first exon remains on the $8q^-$ chromosome and the coding sequences are translocated to chr. $14q^+$[337]. The data further indicate that the first exon of c-myc is rearranged 3′ to 3′ with Ψ_ε gene, into the switch region of the $C_{\alpha 1}$ gene, the prominent site of c-myc translocation seen in mouse plasmacytomas[221]. Nilsen and Maroney[338] have assessed the translational efficiency of c-myc mRNAs in a variety of cell lines carrying normal or translocated c-myc alleles. It is of considerable interest that the mRNA contained in P3HR-1 cells (the transcriptionally active c-myc allele is broken in the first intron and thus the suggested stem-loop structure cannot form) is translated with equal efficiency as normal c-myc mRNAs. This does not support the translational regulation model suggested by Saito et al.[218] (Fig. 2.B).

Dani et al.[339] have shown that the half-life of c-myc mRNA is extremely short, about 10 min. Their finding that the inhibition of protein synthesis in some, but not all cells resulted in the stabilization of this message suggests the possibility that another type of post-transcriptional control of c-myc expression may exist. It is apparent from their data that a protein, itself with a rapid turn-over, might be responsible for exerting such effects.

The molecular analysis of the Burkitt's lymphoma AW-Ramos cell line by Wiman et al.[340] lends further credence to the model of c-myc regulation mediated by 5′-regulatory sequences (Fig. 2.A), and also suggests that the constitutive expression of c-myc is more important than the absolute level of c-myc mRNA.

Experiments by Lachman and Skoultchi[341] add yet another dimension to c-myc regulation. They studied the level of c-myc mRNA in mouse erythroleukemia cells induced to differentiate by dimethyl sulfoxide or hypoxanthine, and observed a biphasic change of c-myc expression in the first 24 h after addition of inducer to the culture. An initial decline of c-myc expression does not require protein synthesis, whereas a restoration of c-myc mRNA levels is dependent upon continued protein synthesis. The observation of an early decline in c-myc expression can be explained by the conventional negative feed-back control model outlined in Fig. 2.A. The restoration of expression may, however, suggest that c-myc expression may also be controlled by a positively acting protein factor.

A stringent negative as well as positive regulation of c-myc expression was also observed in WEHI 231 cells, which, when treated with anti-mouse Ig antiserum cease proliferation by 24 h after completing one round of cell division and become arrested in G_1 phase. This arrest of growth by a specific receptor (sIg)-mediated signal is preceded by a dramatic decline in cytoplasmic levels of c-myc mRNA, which, during the initial phase of treatment, is transiently increased 5- to 10-fold[342].

6 Acknowledgement

Thanks are due to Professor D. Drahovsky for the opportunity to work in his laboratory, his encouraging interest in medical genetics and many stimulating discussions.

Financial support for the original studies reported here was provided by grants from the Deutsche Forschungsgemeinschaft to D.D. and the Deutsche Krebshilfe.

7 References

1. Boveri, T.: Zur Frage der Entstehung maligner Tumoren, Jena, Fischer 1914
2. Graham, F. L., vander Eb, A. J.: Virology *52,* 456 (1973)
3. Shih, C., Shilo, B.-Z., Goldfarb, M. P., Dannenberg, A., Weinberg, R. A.: Proc. Natl. Acad. Sci. USA *76,* 5714 (1979)
4. Cooper, G. M., Okenquist, S., Silverman, L.: Nature *284,* 413 (1980)
5. Krontiris, T. G., Cooper, G. M.: Proc. Natl. Acad. Sci. USA *78,* 1181 (1981)
6. Weinberg, R. A.: Biochem. Biophys. Acta *651,* 25 (1981)
7. Cooper, G. M.: Science *217,* 801 (1982)
8. Rigby, P. W. J.: Nature *290,* 186 (1981)
9. Smith, B. L., Anisowicz, A., Chodosh, L. A., Sager, R.: Proc. Natl. Acad. Sci. USA *79,* 1964 (1982)
10. Sutherland, B. M., Bennett, P. V.: Cancer Res. *44,* 2769 (1984)
11. Blair, D. G., Cooper, C. S., Oskarsson, M. K., Eader, L. A., Vande Woude, G. F.: Science *218,* 1122 (1982)
12. Perucho, M., Goldfarb, M., Shimizu, K., Lama, C., Fogh, J., Wigler, M.: Cell *27,* 467 (1981)
13. Lane, M.-A., Sainten, A., Cooper, G. M.: Proc. Natl. Acad. Sci. USA *78,* 5185 (1981)
14. Lane, M.-A., Sainten, A., Cooper, G. M.: Cell *28,* 873 (1982)
15. Goubin, G., Goldman, D. S., Luce, J., Neiman, P. E., Cooper, G. M.: Nature *302,* 114 (1983)
16. Lane, M.-A., Sainten, A., Doherty, K. M., Cooper, G. M.: Proc. Natl. Acad. Sci. USA *81,* 2227 (1984)
17. Cooper, C. S., Blair, D. G., Oskarsson, M. K., Tainsky, M. A., Eader, L. A., Vande Woude, G. F.: Cancer Res. *44,* 1 (1984)
18. Pulciani, S., Santos, E., Lanver, A. V., Long, L. K., Aaronson, S. A., Barbacid, M.: Nature *300,* 539 (1982)
19. Clarke, M. F., Westin, E., Schmidt, D., Josephs, S. F., Ratner, L., Wong-Staal, F., Gallo, R. C., Reitz, M. S.: Nature *300,* 464 (1984)
20. Okazaki, W., Purchase, H. G., Crittenden, L. B.: Avian Dis. *26,* 553 (1982)
21. Hayward, W. S., Neel, B. G., Astrin, S. M.: Nature *290,* 475 (1981)
22. Neel, B. G., Hayward, W. S., Robinson, H. L., Fang, J., Astrin, S. M.: Cell *23,* 323 (1981)
23. Schubach, W., Groudine, M.: Nature *307,* 702 (1984)
24. Fung, Y.-K. T., Lewis, W. G., Crittenden, L. B., Kung, H.-J.: Cell *33,* 357 (1983)
25. Westaway, D., Payne, G., Varmus, H. E.: Proc. Natl. Acad. Sci. USA *81,* 843 (1984)
26. Cullen, B. R., Comedico, P. T., Ju, G.: Nature *307,* 241 (1984)
27. Steffen, D.: Proc. Natl. Acad. Sci. USA *81,* 2097 (1984)
28. Cuypers, H. T., Selten, G., Quint, W., Zijlstra, M., Maandag, E. R., Boelens, W., Wezenbeek, P. V., Melief, C., Berns, A.: Cell *37,* 141 (1984)
29. Nusse, R., v. Ooyen, A., Cox, D., Fung, Y.-K. T., Varmus, H. E.: Nature *307,* 131 (1984)
30. Peters, G., Brookes, S., Smith, R., Dickson, C.: Cell *33,* 369 (1983)

30a. Peters, G., Kozak, C., Dickson, C.: Mol. Cell. Biol. *4,* 375 (1984)

31. Seiki, M., Eddy, R., Shows, T. B., Yoshida, M.: Nature *309,* 640 (1984)
32. deleted.
33. Sodraski, J. G., Rosen, C. A., Haseltine, W. A.: ibid. *225,* 381 (1984)

33a. Shaw, G. M., Gouda, M. A., Flickinger, G. H., Hahn, B. H., Gallo, R. C., Wong-Staal, F.: Proc. Natl. Acad. Sci. USA *81,* 4544 (1984)

34. Manzari, V., Gallo, R. C., Franchini, G., Westin, E., Cecherini-Nelli, L., Popovic, M., Wong-Staal, F.: ibid. *80,* 11 (1983)
35. Cohen, C.-J., Murphey-Corb, M.: Nature *301,* 129 (1984)
36. Rechavi, G., Givol, D., Canaani, E.: ibid. *300,* 607 (1983)
37. Kuff, E. L., Feenstra, A., Lueders, K., Rechavi, G., Givol, D., Canaani, E.: ibid. *302,* 547 (1983)
38. Canaani, E., Dreazen, O., Klar, A., Rechavi, G., Ram, D., Cohen, J. B., Givol, D.: Proc. Natl. Acad. Sci. USA *80,* 7118 (1983)
39. Cohen, J. B., Unger, T., Rechavi, G., Canaani, E., Givol, D.: Nature *306,* 797 (1983)
40. Bishop, J. M.: Adv. Cancer Res. *37,* 1 (1982)

41. Vennstroem, B., Bishop, J. M.: Cell *28,* 135 (1982)
42. Nunn, M. F., Seeburg, M. H., Moscovici, C., Duesberg, P. H.: Nature *306,* 391 (1983)
43. Leprince, D., Gegonne, A., Coll, J., de Taisne, C., Schneeberger, A., Lagron, C., Stehelin, D.: Nature *306,* 395 (1983)
44. Coll, J., Righi, M., deTaisne, C., Dissons, C., Gegonne, A., Stehelin, D.: EMBO J. *2,* 2189 (1983)
45. Kan, N. C., Flordellis, C. S., Garon, C. F., Duesberg, P. H., Papas, T. H. H.: Proc. Natl. Acad. Sci. USA *80,* 6566 (1983)
46. Jansen, H. W., Rückert, B., Lurz, R., Bister, K.: EMBO J. *2,* 1969 (1983)
47. Jansen, H. W., Kurz, R., Bister, K., Bonner, T. I., Mark, G. E., Rapp, U. R.: Nature *307,* 281 (1984)
48. Kan, N. C., Flordellis, C. S., Mark, G. E., Duesberg, P. H., Papas, T. H.: Science *223,* 813 (1984)
48a. Naharro, G., Robbins, K. C., Reddy, E. R.: ibid. *223,* 63 (1984)
49. Symonds, G., Stubblefield, E., Guyaux, M., Bishop, J. M.: Mol. Cell. Biol. *4,* 1627 (1984)
50. Zabel, B. U., Fournier, R. E. K., Lalley, P. A., Naylor, S. L., Sakaguchi, A. Y.: Proc. Natl. Acad. Sci, USA *81*, 4874 (1984)
51. Iba, H., Takeya, T., Cross, F. R., Hanafusa, T., Hanafusa, H.: ibid. *81,* 4424 (1984)
52. Miller, A. D., Curran, T., Verma, I. M.: Cell *36,* 51 (1984)
53. Verma, I. M.: Nature *308,* 317 (1984)
54. Wang, J. Y. J.: ibid. *304,* 400 (1983)
55. Duesberg, P. H.: ibid. *304,* 219 (1983)
56. Bishop, J. M.: Ann. Rev. Biochem. *52,* 301 (1983)
57. Mark, G. E., Rapp, U. R.: Science *224,* 285 (1984)
58. Sutrave, P., Bonner, T. I., Rapp, U. R., Jansen, H. W., Patschinsky, T., Bister, K.: Nature *309,* 85 (1984)
59. Groffen, J., Heisterkamp, N., Skibuya, M., Hanafusa, H., Stephenson, J. R.: Virology *125,* 480 (1983)
60. Kamps, M. P., Taylor, S. S., Sefton, B. M.: Nature *310,* 589 (1984)
60a. Groffen, J., Heisterkamp, N., Reynolds, F. H., Stephenson, J. R.: ibid. *304,* 167 (1983)
61. Hoffmann, F. M., Fresco, L. D., Hoffmann-Falk, H., Shilo, B.-Z.: Cell *35,* 393 (1983)
62. Lev, Z., Leibovitz, N., Seger, O., Shilo, B.-Z.: Mol. Cell. Biol. *4,* 982 (1984)
63. DeFeo-Jones, D., Scolnick, F., Koller, R., Dher, R.: Nature *306,* 707 (1983)
64. Gallwitz, D., Dorath, C., Sander, C.: ibid. *306,* 704 (1983)
65. Dhar, R., Nieto, A., Koller, R., DeFeo-Jones, D., Scolnick, E. M.: Nuck. Acids Res. *12,* 3611 (1984)
66. Powers, S., Kataoka, T., Fasano, O., Goldfarb, M., Strathern, J., Broach, J., Wigler, M.: Cell *36,* 607 (1984)
66a. Coffin, J. M., Varmus, H. E., Bishop, J. M., Essex, M., Hardy, W. D., Martin, G. S., Rosenberg, N. E., Scolnick, E. M., Weinberg, R. A., Vogt, P. K.: J. Virol. *40,* 953 (1981)
67. Yunis, J. J.: Science *221,* 227 (1983)
68. Sandberg, A. A.: Cancer Genet. Cytogenet. *8,* 277 (1983)
69. Gilbert, F.: J. Natl. Cancer Inst. *71,* 1107 (1983)
70. LeBeau, M. M., Rowley, J. D.: Nature *308,* 607 (1984)
71. Yunis, J. J., Bloomfield, C. D., Ensrud, K.: N. Engl. J. Med. *305,* 135 (1981)
72. Yunis, J. J., Oken, M. M., Kaplan, M. E., Ensrud, K. M., Howe, R. R., Theologides, A.: N. Engl. J. Med. *307,* 1231 (1982)
73. Rowley, J. D.: Clin. Haematol. *9,* 55 (1980)
74. deKlein, A., van Kessel, A. G., Grosveld, G., Bartram, C. R., Hagemeijer, A., Bootsma, D., Spurr, N. K., Heisterkamp, N., Groffen, J., Stephenson, J. R.: Nature *300,* 765 (1982)
75. Rabbitts, T. H., Forster, A., Matthews, J. G.: Mol. Biol. Med. *1,* 11 (1983)
76. Bartram, C. R., deKlein, A., Hagemeijer, A., Grosveld, G., Heisterkamp, N., Groffen, J.: Blood *63,* 223 (1984)
77. Morton, C., Taub, R., Diamond, A., Lane, M. A., Cooper, G. M., Leder, P.: Science *223,* 173 (1984)
78. deMartinville, B., Cunningham, J. M., Murray, M. J., Francke, U.: Nucl. Acids Res. *11,* 5267 (1983)

79. Solomon, E., Goodfellow, P.: Nature *306*, 123 (1984)
80. Rabin, M., Watson, M., Barker, P. E., Ryan, J., Breg, W. R., Ruddle, F. H.: Cytogenet. Cell Genet. *38*, 70 (1984)
81. Malcolm, S., Barton, P., Murphy, C., Fergusson-Smith, M. A., Bentley, D. L., Rabbitts, T. H.: Proc. Natl. Acad. Sci. USA *79*, 4957 (1982)
82. Schwab, M., Varmus, H. E., Bishop, J. M., Grzeschik, K.-H., Naylor, S. L., Sakaguchi, A. Y., Brodeur, G., Trent, J.: Nature *308*, 288 (1984)
83. Bonner, T., O'Brien, S. J., Nash, N. G., Rapp, U. R., Morton, C. C., Leder, P.: Science *223*, 71 (1984)
84. Roussel, M. F., Sherr, C. J., Barker, P. E., Ruddle, F. H.: J. Virol. *48*, 770 (1983)
85. McBride, O. W., Swan, D. C., Tronick, S. R., Gol, R., Klimanis, D., Moore, D. E., Aaronson, S. A.: Nucl. Acids Res. *11*, 8221 (1983)
86. Sakaguchi, A. Y., Zabel, B. U., Grzeschik, K.-H., Law, M. L., Ellis, R. W., Scolnick, E. M., Naylor, S. L.: Mol. Cell. Biol. *4*, 989 (1984)
87. Harper, M. E., Franchini, G., Love, J., Simon, M. I., Gallo, R. C., Wong-Staal, F.: Nature *304*, 169 (1983)
87a. Zabel, B. U., Naylor, S. L., Grzeschik, K.-H., Sakaguchi, A. Y.: Somatic Cell. Mol. Genet. *10*, 105 (1984)
88. Spurr, N. K., Solomon, E., Jannson, M., Sheer, D., Goodfellow, P. N., Bodmer, W. F., Vennstrom, B.: EMBO J. *3*, 159 (1984)
89. Neel, B. G., Jhanwar, S. C., Chaganti, R. S. K., Hayward, W. S.: Proc. Natl. Acad. Sci. USA *79*, 7842 (1982)
90. Jhanwar, S. C., Neel, B. G., Hayward, W. S., Chaganti, R. S. K.: Cytogenet. Cell Genet. *38*, 73 (1984)
91. deMartinville, B., Giacalone, J., Shih, C., Weinberg, R. A., Francke, U.: Science *219*, 498 (1983)
92. deTaisne, C., Gegonne, A., Stehelin, D., Bernheim, A., Berger, A.: Nature *310*, 581 (1984)
93. Croce, C. M., Shander, M., Martinis, J., Cicurel, L., D'Amona, G. G., Dolby, T. W., Koprowsky, H.: Proc. Natl. Acad. Sci. USA *76*, 3416 (1979)
94. Hobart, M. J., Rabbitts, T. H., Goodfellow, P. N., Solomon, E., Chambers, S., Spurr, N., Povey, S.: Am. J. Human Genet. *45*, 331 (1981)
95. Kisch, I. R., Morton, C. C., Nakahara, K., Leder, P.: Science *216*, 301 (1982)
96. Dayton, A. I., Selden, J. R., Laws, G., Dorney, D. J., Finan, J., Triputti, P., Emanuell, B. S., Rovera, G., Nowell, P. C., Corce, C. M.: Proc. Natl. Acad. Sci. USA *81*, 4495 (1984)
97. LeBeau, M. M., Wsetbrook, C. A., Diaz, M. O., Rowley, J. D.: Nature *312*, 70 (1984)
98. Erikson, J., Finan, J., Nowell, P. C., Croce, C. M.: Proc. Natl. Acad. Sci. USA *79*, 5611 (1982)
99. Rabbitts, T. H., Forster, A., Matthews, J.: Mol. Biol. Med. *1*, 11 (1983)
100. O'Brien, S. J., Nashe W. G., Goodwin, J. L., Lowy, D. R., Chang, E. H.: Nature *302*, 839 (1983)
101. Sheiness, D., Gardinier, M.: Mol. Cell. Biol. *4*, 1206 (1984)
102. Gonda, T. J., Metcalf, D.: Nature *310*, 249 (1984)
103. Westin, E. H., Wong-Staal, F., Gelmann, E. P., Dalla-Favera, R., Papas, T. S., Lautenberger, J. A., Eva, A., Reddy, E. P., Tronick, S. R., Aaronson, S. A., Gallo, R. C.: Natl. Acad. Sci. USA *79*, 2490 (1982)
104. Westin, E. H., Gallo, R. C., Avya, S. K., Eva, A., Souza, L. M., Baluda M. A., Aaronson, S. A., Wong-Staal, F.: ibid. *79*, 2194 (1982)
105. Craig, R. W., Bloch, A.: Cancer Res. *44*, 442 (1984)
106. Rosson, D., Tereba, A.: Cancer Res. *43*, 3912 (1983)
107. Goyette, M., Petropoulos, C. J., Shank, P. R., Fausto, N.: Mol. Cell. Biol. *4*, 1493 (1984)
108. Makino, R., Hayashi, K., Sato, S., Sugimura, T.: Biochem. Biophys. Res. Comm. *119*, 1096 (1984)
109. Makino, R., Hayashi, K., Sugimura, T.: Nature *310*, 697 (1984)
110. Sheiness, D., Gardinier, M.: Mol. Cell Biol. *4*, 1206 (1984)
111. Campisi, J., Gray, H. E., Pardee, A. B., Dean, M., Sonenshein, G. E.: Cell *36*, 241 (1984)
112. Jonak, G. J., Knight, E.: Proc. Natl. Acad. Sci. USA *81*, 1747 (1984)
113. Reitsma, P. H., Rothberg, P. G., Astrin, S. M., Trial, J., Bar-Shavit, Z., Hall, A., Teitelbaum, S. L., Kahn, A. J.: Nature *306*, 492 (1983)

114. Slamon, D. J., deKernion, J. B., Verma, J. M., Cline, M. J.: Science *224,* 256 (1984)
115. Müller, R., Slamon, D. J., Trembley, J. M., Cline, M. J., Verma, J. M.: Nature *299,* 640 (1982)
116. Boehm, T. L. J., Kaul, S., Kornhuber, B., Drahovsky, D.: Verh. Deutsche Krebsges. *5,* 154 (1984)
117. Eva, A., Robbins, K. C., Andersen, P. R., Srinivasan, A., Tronick, S. R., Reddy, E. P., Ellmore, N. W., Galen, A. T., Lautenberger, J. A., Papas, T. S., Westin, E. H., Wong-Staal, F., Gallo, R. C., Aaronson, S. A.: Nature *295,* 116 (1982)
118. Müller, R., Trembley, J. M.: Adamson, E. D., Verma, I. M.: ibid. *304,* 454 (1983)
119. Alitalo, K., Schwab, M., Liu, C. C., Varmus, H. E., Bishop, J. M.: Proc. Natl. Acad. Sci. USA *80,* 1707 (1983)
120. Alitalo, K., Winquist, R., Liu, C. C., de la Chapelle, A., Schwab, M., Bishop, J. M.: ibid. *81,* 4534 (1984)
121. Schwab, M., Alitalo, K., Varmus, H. E., Bishop, J. M., George, D.: Nature *309,* 497 (1983)
122. Little, C. D., Nan, M. M., Carney, D. N., Gazdar, A. F., Mina, J. D.: Nature *306,* 194 (1984)
123. Brodeur, G., Seeger, C., Schwab, M., Varmus, H. E., Bishop, J. M.: Science *224,* 1121 (1984)
124. Schwab, M., Varmus, H. E., Bishop, J. M., Grzeschik, K.-H., Naylor, S., Sakaguchi, A., Brodeur, G., Trent, J.: Nature *308,* 288 (1984)
125. Schwab, M., Ellison, J., Busch, M., Rosenau, W., Varmus, H. E., Bishop, J. M.: Proc. Natl. Acad. Sci. USA *81,* 4940 (1984)
126. Collins, S., Groudine, M.: Nature *298,* 679 (1982)
127. Dalla-Favera, R., Wong-Staal, F., Gallo, R. C.: Nature *299,* 61 (1982)
128. Kozbor, D., Croce, C. M.: Cancer Res. *44,* 438 (1984)
129. Pelici, P.-G., Lanfrancone, L., Varmus, H. E., Bishop, J. M.: Science *224,* 1117 (1984)
130. Collins, S. J., Groudine, M. T.: Proc. Natl. Acad. Sci. USA *80,* 4813 (1983)
131. Selden, J. R., Emanuel, B. S., Wang, E., Cannizarro, L., Palumbo, A., Erikson, J., Nowell, P. C., Rovera, G., Croce, C. M.: ibid. *80,* 7289 (1983)
132. Nowell, P., Finan, J., Dalla-favera, R., Gallo, R. C., Ar-Rushdi, A., Romanczuk, H., Selden, J. R., Emanuel, B. S., Rovera, G., Croce, C. M.: Nature *306,* 494 (1983)
133. Fasano, O., Aldrich, T., Tamanoi, F., Taparowsky, E., Furth, M., Wigler, M.: Proc. Natl. Acad. Sci. USA *81,* 4771 (1984)
134. Feinberg, A. P., Vogelstein, B., Droller, M.-J., Baylin, S. B., Nelkin, B. D.: Science *220,* 1175 (1983)
135. Souyri, M., Fleissner, E.: Proc. Natl. Acad. Sci. USA *80,* 6676 (1983)
136. Albino, A. P., LeStrange, R., Oliff, A. I., Furth, M. E., Old, L. J.: Nature *308,* 69 (1984)
137. Gambke, C., Signer, E., Moroni, C.: Nature *307,* 476 (1984)
138. Murray, M. J., Cunningham, J. M., Parada, L. F., Dantry, F., Lebowitz, P., Weinberg, R. A.: Cell *33,* 749 (1983)
139. Hall, A., Marshall, C. J., Spurr, N. K., Weiss, R. A.: Nature *303,* 396 (1983)
140. Taparowsky, E., Suard, Y., Fasano, O., Shimizu, K., Goldfarb, M., Wigler, M.: ibid. *300,* 762 (1982)
141. Yuasa, Y., Gol., R. A., Chang, A., Chiu, I.-M., Reddy, E. P., Tronick, S. R., Aaronson, S. A.: Proc. Natl. Acad. Sci. USA *81,* 3670 (1984)
142. Tabin, C. J., Bradley, S. M., Bargmann, C. I., Weinberg, R. A., Papageorge, A.-G., Scolnick, E. M., Dhar, R., Lowy, D. R., Chang, E. H.: Nature *300,* 143 (1982)
143. Yuasa, Y., Srivastava, S. K., Dunn, C. Y., Rhim, J. S., Reddy, E. P., Aaronson, S. A.: ibid. *303,* 775 (1983)
144. Sekiya, T., Fushimi, M., Hori, H., Hirohashi, S., Nishimura, S., Sugimura, T.: Proc. Natl. Acad. Sci. USA *81,* 4777 (1984)
145. Taparowsky, E., Shimizu, K., Goldfarb, M., Wigler, M.: Cell *34,* 381 (1983)
146. McCoy, M. S., Toole, J. J., Cunningham, J. M., Chang, E. H., Lowy, D. R., Weinberg, R. A.: Nature *302,* 79 (1983)
147. Pulciani, S., Santos, E., Lanver, A. V., Long, L. K., Aaronson, S. A., Barbacid, M.: ibid. *300,* 539 (1982)
148. Nakano, H., Yamamoto, F., Neville, C., Evans, D., Mizumo, T., Perucho, M.: Proc. Natl. Acad. Sci. USA *81,* 71 (1984)
149. Capon, D. J., Seeburg, P. H., McGrath, J. P., Hayflick, J. S., Edman, U., Levinson, A. D., Goeddel, D. V.: Nature *304,* 507 (1983)

150. McCoy, M. S., Bargmann, C. I., Weinberg, R. A.: Mol. Cell. Biol. *4,* 1577 (1984)
151. Santos, E., Martin-Zanca, D., Reddy, E. P., Pierotti, M. A., Della-Porta, G., Barbacid, M.: Science *223,* 661 (1984)
152. Feig, L. A., Bast, R. C., Knapp, R. C. Cooper, G. M.: Science *223,* 698 (1984)
153. O'Toole, C. M., Povey, S., Hepburn, P., Franks, L. M.: Nature *301,* 429 (1983)
154. Mushinksi, J. F., Potter, M., Bauer, S. R., Reddy, E. P.: Science *220,* 795 (1983)
155. Robbins, K. C., Antoniades, H. N., Devare, S. G., Hunkapillar, M. W., Aaronson, S. A.: Nature *305,* 605 (1983)
156. Doolittle, R. F., Hunkapillar, M. W., Hood, L. E., Devare, S. G., Robbins, K. C., Aaronson, S. A., Antoniades, H. N.: Science *221,* 275 (1983)
157. Chiu, I.-M., Reddy, E. P., Givol, D., Robbins, K. C., Tronick, S. R., Aaronson, S. A.: Cell *37,* 123 (1984)
158. Niman, H. L.: Nature *307,* 180 (1984)
159. Stiles, C. D.: Cell *33,* 653 (1983)
159a. Hedlin, C. H., Westmark, B., Wasteson, A.: Proc. Natl. Acad. Sci. USA *78,* 3664 (1981)
160. Ek, B., Westermark, B., Wasteson, A., Heldin, C.-H.: Nature *295,* 419 (1982)
161. Ulsh, L. S., Shih, T. Y.: Mol. Cell. Biol. *4,* 1647 (1984)
162. Sweet, R. W., Yokoyama, S., Kamata, T., Feramisco, J. R., Rosenberg, M., Gross, M.: Nature *311,* 273 (1984)
163. McGrath, J. P., Capon, D. J., Goeddel, D. V., Levinson, A. D.: ibid. *310,* 644 (1984)
164. Finkel, T., Der, C. J., Cooper, G. M.: Cell *37,* 151 (1984)
165. Der, C. J., Cooper, G. M.: Cell *32,* 201 (1983)
166. Pincus, M. R., van Renswoude, J., Harford, J. B., Chang, E. H., Carty, R. P., Klausner, R. D.: Proc. Natl. Acad. Sci. USA *80,* 5253 (1983)
167. Willumsen, B. M., Christensen, A., Hubbert, N. L., Papageorge, A. G., Lowy, D. R.: Nature *310,* 583 (1984)
168. Giallongo, A., Appella, E., Ricciardi, R., Rovera, G., Croce, C. M.: Science *222,* 430 (1983)
169. Downward, J., Yarden, Y., Mayes, E., Scrace, G., Totty, N., Stockwell, P., Ullrich, A., Schlessinger, J., Waterfield, M. D.: Nature *307,* 521 (1984)
170. Shimizu, N., Behzadian, M. A., Shimizu, Y.: Proc. Natl. Acad. Sci. USA *77,* 3600 (1980)
171. Davies, R. L., Grosse, V. A., Kucherlapati, R., Bothwell, M.: ibid. *77,* 4188 (1980)
172. Hunter, T.: Nature *311,* 414 (1984)
173. Cooper, G. M., Neiman, P. E.: ibid. *287,* 656 (1980)
174. Cooper, G. M., Neiman, P. E.: ibid. *292,* 857 (1981)
175. Lane, M.-A., Neary, D., Cooper, G. M.: ibid. *300,* 659 (1982)
176. Murray, M. J., Cunningham, J. M., Parad. L. F., Dantry, F., Lebowitz, P., Weinberg, R. A.: Cell *33,* 749 (1983)
177. Land, H., Parada, L. F., Weinberg, R. A.: Nature *304,* 596 (1983)
178. Newbold, R. F., Overell, R. W.: ibid. *304,* 648 (1983)
179. Ruley, H. E.: ibid. *304,* 602 (1983)
180. Jariwalla, R. J., Aurelian, L., Ts'o, P. O. P.: Proc. Natl. Acad. Sci. USA *80,* 5902 (1983)
181. Donoghue, D. J., Anderson, C., Hunter, T., Kaplan, P. L.: Nature *308,* 748 (1984)
182. Asselin, C., Gélinas, C., Branton, P. E., Bastin, M.: Mol. Cell. Biol. *4,* 755 (1984)
183. Donner, P., Greiser-Wilke, I., Moelling, K.: Nature *296,* 262 (1982)
184. Abrakas, H. D., Rohrschneider, L. R., Eisenman, R. N.: Cell *29,* 427 (1982)
185. Ito, Y., Spurr, N., Dulbecco, R.: Proc. Natl. Acad. Sci. USA *74,* 1259 (1977)
186. Feldman, L. T., Nevins, J. P.: Mol. Cell. Biol. *3,* 829 (1983)
187. Ralston, R., Bishop, J. M.: Nature *306,* 803 (1983)
188. Pereira-Smith, O. M., Smith, J. R.: Science *221,* 964 (1983)
189. Huschtscha, I. T., Holliday, R.: J. Cell Sci. *63,* 77 (1983)
190. Stanbridge, E. J., Willkinson, J. E.: Int. J. Cancer *26,* 1 (1980)
191. Heldin, C.-H., Westermark, B.: Cell *37,* 9 (1984)
192. Diringer, H., Friis, R. R.: Cancer Res. *37,* 2978 (1977)
193. Nishizuka, Y.: Nature *308,* 693 (1984)
194. Streb, H., Irvine, R. F., Berridge, M. J., Schulz, I.: ibid. *306,* 67 (1983)
195. Sugimoto, Y., Whitman, M., Cantley, L. C., Erikson, R. L.: Proc. Natl. Acad. Sci. USA *81,* 2117 (1984)

196. Gilman, A.: Cell *36,* 577 (1984)
Ref. 197–200 were deleted during preparation of manuscript
201. Erikson, J., Finan, J., Tsujimoto, Y., Nowell, P. C., Croce, C. M.: Proc. Natl. Acad. Sci. USA *81,* 4144 (1984)
202. Tsujimoto, Y., Yunis, J. J., Ouorato, M., Showe, L., Erikson, J., Nowell, P. C., Croce, C. M.: Science *224,* 1403 (1984)
203. Korsmeyer, S. J., Hieter, P. A., Ravetech, J. V., Poplack, D. G., Waldmann, T. A., Leder, P.: Proc. Natl. Acad. Sci. USA *78,* 7096 (1981)
204. deleted
205. Bakhshi, A., Minowada, J., Arnold, A., Cossmann, J., Jansen, J. P., Whang-Peng, J., Waldmann, T. A., Korsmeyer, S. J.: N. Engl. J. Med. *309,* 826 (1983)
206. Ford, A. M., Molgaard, H. V., Greaves, M. F., Gould, H. J.: EMBO J. *2,* 997 (1983)
207. Clearly, M. L., Chao, J., Wanke, R., Sklar, J.: Proc. Natl. Acad. Sci. USA *81,* 593 (1984)
208. Clearly, M. L., Warnke, R., Sklar, J.: N. Engl. J. Med. *310,* 477 (1984)
209. Feroni, L., Catovsky, D., Rabbitts, T. H., Luzzatto, L.: Mol. Biol. Med. *2,* 63 (1984)
210. Toyonaga, B., Yanagi, Y., Suciu-Foca, N., Minden, M., Mak, T. W.: Nature *311,* 385 (1984)
211. Bartram, C. R., deKlein, A., Hagemeijer, A., van Agthoven, T., van Kessel, A. G., Bootsma, D., Grosveld, G., Ferguson-Smith, M. A., Davies, T., Stone, M., Heisterkamp, N., Stephenson, J. R., Groffen, J.: Nature *306,* 277 (1983)
212. Mintz, U., Vardiman, J., Golomb, H. N., Rowley, J. D.: Cancer *43,* 411 (1979)
213. Heisterkamp, N., Stephenson, J. R., Groffen, J., Hansen, P. F., deKlein, A., Bartram, C. R., Grosveld, G.: Nature *306,* 239 (1983)
214. Groffen, J., Stephenson, J. R., Heisterkamp, N., deKlein, A., Bartram, C. R., Grosveld, G.: Cell *36,* 93 (1984)
215. Colby, W. W., Chen, E. Y., Smith, D. H., Levinson, A. D.: Nature *301,* 722 (1983)
216. Watson, D. K., Psallidopoulos, M. C., Samual, K. P., Dalla-Favera, R., Papas, T. S.: Proc. Natl. Acad. Sci. USA *80,* 3642 (1983)
217. Watt, R., Stanton, L. W., Marcu, K. B., Gallo, R. C., Croce, C. M., Rovera, G.: Nature *303,* 725 (1983)
218. Saito, H., Hayday, A. C., Wiman, K., Hayward, W. S., Tonegawa, S.: Proc. Natl. Acad. Sci. USA *80,* 7476 (1983)
219. Gazin, C., de Dinechin, S. D., Hampe, A., Masson, J.-M., Martin, P., Stehelin, D., Galibert, F.: EMBO J. *3,* 383 (1984)
220. Hayday, A. C., Gillies, S. D., Saito, H., Ward, C., Wiman, K., Hayward, W. S., Tonegawa, S.: Nature *307,* 334 (1984)
221. Adams, J. M., Gerondakis, S., Webb, E., Corcoran, L. M., Cory, S.: Proc. Natl. Acad. Sci. USA *80,* 1982 (1983)
222. Stanton, L. W., Watt, R., Marcu, K. B.: Nature *303,* 401 (1983)
223. Stanton, L. W., Fahrlander, P. D., Tesser, P. M., Marcu, K. B.: ibid. *310,* 423 (1984)
224. Maguire, R. T., Robins, T. S., Thorgeirsson, S. S., Heilman, C. A.: Proc. Natl. Acad. Sci. USA *80,* 1947 (1983)
225. Nishikura, K., Ar-Rushdi, A., Erikson, J., Watt, R., Rovera, G., Croce, C. M.: ibid. *80,* 4822 (1983)
226. Ar-Rushdi, A., Nishikura, K., Erikson, J., Watt, R., Rovera, G., Croce, C. M.: Science *222,* 390 (1983)
227. Croce, C. M., Erikson, J., Ar-Rushdi, A., Aden, D., Nishikura, K.: Proc. Natl. Acad. Sci. USA *81,* 3170 (1984)
228. Nishikura, K., Ar-Rushdi, A., Erikson, J., Defesus, E., Dugan, D., Croce, C. M.: Science *224,* 399 (1984)
229. Taub, R., Moulding, C., Battey, J., Murphy, W., Vasicek, T., Lenoir, G., Leder, P.: Cell *36,* 339 (1984)
230. Tonegawa, S.: Nature *302,* 575 (1983)
231. Rabbitts, T. H., Forster, A., Hamlyn, P., Baer, R.: ibid. *309,* 592 (1984)
232. Leder, P., Battey, J., Lenoir, G., Moulding, C., Murphy, W., Potter, H., Stewart, T., Taub, R.: Science *222,* 765 (1983)
233. Hamlyn, P. H., Rabbitts, T. H.: Nature *304,* 135 (1983)
234. Rabbitts, T. H., Forster, A., Bear, R., Hamlyn, P. H.: ibid. *306,* 806 (1983)

235. Erikson, J., Al-Rushdi, A., Drwinga, H. L., Nowell, P. C., Croce, C. M.: Proc. Natl. Acad. Sci. USA *80,* 820 (1983)
236. Adams, J. M., Gerondakis, S., Webb, E., Corcorau, L. M., Cory, S.: ibid. *80,* 1982 (1983)
237. Taub, R., Kirsch, I., Morton, C., Lenoir, G., Swan, D., Tronick, S., Aaronson, S. A., Leder, P.: ibid. *79,* 7837 (1982)
238. Battey, J., Moulding, C., Taub, R., Murphy, W., Stewart, T., Potter, H., Lenoir, G., Leder, P.: Cell *34,* 779 (1983)
239. Gelman, E. P., Psallidopoulos, M. C., Papas, T. S., Dalla-Favera, R.: Nature *306,* 799 (1983)
240. Dalla-Favera, R., Martinotti, S., Gallo, R. C., Erikson, J., Croce, C. M.: Science *219,* 963 (1983)
241. Emanuel, B. S., Selden, J. R., Chaganti, R. S. K., Jhanwar, S., Nowell, P. S., Croce, C. M.: Proc. Natl. Acad. Sci. USA *81,* 2444 (1984)
242. Erikson, J., Nishikura, K., Al-Rushdi, A., Finan, J., Emanuel, B., Lenoir, G., Nowell, P. C., Croce, C. M.: ibid. *80,* 7581 (1983)
243. Taub, R., Kelly, K., Battey, J., Latt, S., Lenoir, G., Tantravahi, U., Tu, Z., Leder, P.: Cell *37,* 511 (1984)
244. Hollis, G. F., Mitchell, K. F., Battey, J., Potter, H., Taub, R., Lenoir, G. M., Leder, P.: Nature *307,* 752 (1984)
245. Croce, C. M., Thierfelder, W., Erikson, J., Nishikura, K., Finan, J., Lenoir, G., Nowell, P. C.: Proc. Natl. Acad. Sci. USA *80,* 6922 (1983)
246. Dalla-Favera, R., Beregni, M., Erikson, I., Patterson, D., Gallo, R. C., Croce, C. M.: ibid. *79,* 7824 (1982)
247. Kelly, K., Cochran, B. H., Stiles, C. D., Leder, P.: Cell *36,* 603 (1983)
248. Boehm, T. L. J., Drahovsky, D.: J. Natl. Cancer Inst. *71,* 429 (1983)
249. Hoffman, R. M.: Biochim. Biophys. Acta *738,* 49 (1984)
250. Riggs, A. D., Jones, P.: Adv. Cancer Res. *40,* 1 (1983)
251. Knudson, A. G.: Proc. Natl. Acad. Sci. USA *68,* 820 (1971)
252. Yunis, J. J., Ramsay, N.: Am. J. Dis. Child. *132,* 162 (1978)
253. Balabau, G., Gilbert, F., Nichols, W., Meadows, A., Shields, J.: Cancer Genet. Cytogenet. *6,* 213 (1982)
254. Sparkes, R. S., Murphree, A. L., Lingua, R. W., Sparkes, M. C., Field, L. L., Funderburk, S. J., Benedict, W. F.: Science *219,* 971 (1983)
255. deMars, R., 23rd Ann. Symp. Fundamental Cancer Res., 1969; p. 105, William and Wilkins, Baltimore, USA, 1970
256. Benedict, W. F., Murphree, A. L., Banerjee, A., Spina, C. A., Sparkes, M. C., Sparkes, R. S.: Science *219,* 973 (1983)
257. Godbout, R., Dryja, T. P., Squire, J. Gallie, B. L., Phillips, R. A.: Nature *304,* 451 (1983)
258. Cavenee, W. K., Dryja, T. P., Phillips, R. A., Benedict, W. F., Godbout, R., Gallie, B. L., Murphree, A. L., Strong, L. C., White, R. L.: ibid. *305,* 779 (1983)
259. Dryja, T. P., Cavenee, W., White, R. L., Rapaport, J. M., Petersen, R., Albert, D. M., Bruns, G. A. P.: N. Engl. J. Med. *310,* 550 (1984)
260. Murphree, A. L., Benedict, W. F.: Science *223,* 1028 (1984)
261. Comings, D. E.: Proc. Natl. Acad. Sci. USA *70,* 3324 (1973)
262. Knudson, A. G.: Paediat. Res. *10,* 513 (1976)
263. Riccardi, V. M., Sujansky, E., Smith, A. C., Francke, U.: Pediatrics *61,* 604 (1978)
264. de Martinville, B., Francke, U.: Nature *305,* 641 (1983)
265. Huerre, C., Depoisse, S., Gilgenkrantz, S., Lenoir, G. M., Junien, C.: ibid. *305,* 638 (1983)
266. Colburn, N. H., Lerman, M. I., Hegamyer, G. A., Gindhart, T. D.: Mol. Cell. Biol. *5*, 890 (1985)
267. Kozak, C. A., Strauss, P. G., Tschilis, P. N.: Mol. Cell. Biol. *5*, 894 (1985)
268. Selten, G., Cuypers, H. T., Zijlstra, M., Melief, C., Berns, A.: EMBO J. *3*, 3215 (1984)
269. Ooyen, A. v., Nusse, R.: Cell *39*, 233 (1984)
270. Debuire, B., Henry, C., Benaissa, M., Biserte, G.: Science *244*, 1456 (1984)
271. Kahn, P., Adkins, B., Beug, H., Graf, T.: Proc. Natl. Acad. Sci. USA *81*, 7122 (1984)
272. Schechter, A. L., Stern, D. F., Vaidyanathan, L., Decker, S. J., Drebin, J. A., Greene, M. I., Weinberg, R. A.: Nature *312*, 513 (1984)
273. Drebin, J. A., Stern, D. F., Link, V. C., Weinberg, R. A., Greene, M. I.: Nature *312*, 545 (1984)

274. Yunis, J. J., Soreng, A. L.: Science *226*, 1199 (1984)
275. De Braekeler, M., Smith, B., Lin, C. C.: Hum. Genet. *69*, 112 (1985)
276. Barker, P. E., Rabin, M., Watson, M., Breg, W. R., Ruddle, F. M., Verma, I. M.: Proc. Natl. Acad. Sci. USA *81*, 5826 (1984)
277. Parker, R. C., Mardon, G., Lebo, R. V., Varmus, H. E., Bishop, J. M.: Mol. Cell. Biol. *5*, 831 (1985)
278. Collins, M. K. L., Goodfellow, P. N., Spurr, N. K., Solomon, E., Tanigawa, G., Tonegawa, S., Owen, M. J.: Nature *314*, 273 (1985)
279. Barker, P. E., Ruddle, F. H., Royer, H.-D., Acuto, O., Reinherz, E. L.: Science *226*, 348 (1984)
279a. Collins, M. K. L., Goodfellow, P. M., Dunne, M. J., Spurr, N. K., Solomon, E., Owen, M. J.: EMBO J. *3*, 2349 (1984)
280. Pfeifer-Ohlsson, S., Goustin, A. S., Rydnert, J., Wahlström, T., Bjersing, L., Stehelin, D., Ohlsson, R.: Cell *38*, 585 (1984)
281. Einat, M., Resnitzky, D., Kimchi, A.: Nature *313*, 597 (1985)
282. Müller, R., Curran, T., Müller, D., Guilbert, L.: Nature *314*, 546 (1985)
283. Müller, R., Müller, D., Guilbert, L.: EMBO J. *3*, 1887 (1984)
284. Kruijer, W., Cooper, J. A., Hunter, T., Verma, I. M.: Nature *312*, 711 (1984)
285. Coshran, B. H., Zullo, J., Verma, I. M., Stiles, C. D.: Science *226*, 1080 (1984)
286. Greenberg, M. E., Ziff, E. B.: Nature *311*, 433 (1984)
287. Armelin, H. A., Armelin, M. C. S., Kelly, K., Stewart, T., Leder, P., Cochran, B. H., Stiles, C. D.: Nature *310*, 655 (1984)
288. Müller, H., Bravo, R., Burckhardt, J.: Nature *312*, 716 (1984)
289. Kohl, N. E., Gee, C. E., Alt, F. W.: Science *226*, 1335 (1984)
290. Thiele, C. J., Reynolds, C. P., Israel, M. A.: Nature *313*, 404 (1985)
291. Taya, Y., Hosogai, K., Hirohashi, S., Shimosato, Y., Tsuchiya, R., Tsuchida, N., Fushimi, M., Sekiya, T., Nishimura, S.: EMBO J. *3*, 2943 (1984)
292. Shibuya, M., Yokota, J., Ueyama, Y.: Mol. Cell. Biol. *5*, 414 (1985)
293. Persson, H., Leder, P.: Science *225*, 718 (1984)
294. Persson, J., Henninghausen, L., Taub, R., DeGrado, W., Leder, P.: Science *225*, 687 (1984)
295. Watt, R. A., Shatzman, A. R., Rosenberg, M.: Mol. Cell. Biol. *5*, 448 (1985)
296. Ramsay, G., Evan, G. I., Bishop, J. M.: Proc. Natl. Acad. Sci. USA *81*, 7742 (1984)
297. Hann, S. R., Eisenman, R. N.: Mol. Cell. Biol. *4*, 2486 (1984)
298. Manne, V., Yamazaki, S., Kung, H.-F.: Proc. Natl. Acad. Sci. USA *81*, 6953 (1984)
299. Gibbs, J. B., Sigal, I. S., Poe, M., Scolnick, E. M.: Proc. Natl. Acad. Sci. USA *81*, 5704 (1984)
300. Manne, Y., Bekeshi, E., Kung, H.-F.: Proc. Natl. Acad. Sci. USA *82*, 376 (1985)
301. Stacey, D. W., Kung, H.-F.: Nature *310*, 508 (1984)
302. Feramisco, J. R., Gross, M., Kamata, T., Rosenberg, M., Sweet, R. W.: Cell *38*, 109 (1984)
303. Feramisco, J. R., Clark, R., Wong, G., Arnheim, N., Milley, R., McCormick, F.: Nature *314*, 639 (1985)
304. Mulcahy, L. S., Smith, M. R., Stacey, D. W.: Nature *313*, 241 (1985)
305. Schejter, E. D., Shilo, B.-Z.: EMBO J. *4*, 407 (1985)
306. Willumsen, B. M., Norris, K., Papageorge, A. G., Hubbert, N. L., Lowy, D. R.: EMBO J. *3*, 2581 (1984)
307. Srivastava, S. K., Yuasa, Y., Reynolds, S. H., Aaronson, S. A.: Proc. Natl. Acad. Sci. USA *82*, 38 (1985)
308. Hand, P. H., Thor, A., Wunderlich, D., Murano, R., Caruso, A., Schlom, J.: Proc. Natl. Acad. Sci. USA *81*, 5227 (1984)
309. Santos, E., Zanca, D. M., Reddy, E. P., Pierotti, M. A., Porta, G. D., Barbacid, M.: Science *223*, 661 (1984)
310. Kraus, M. H., Yuasa, Y., Aaronson, S. A.: Proc. Natl. Acad. Sci. USA *81*, 5384 (1984)
311. Tainsky, M. A., Cooper, C. S., Giovanella, B. C., Vande Woude, G. F.: Science *225*, 643 (1984)
312. Vousden, K. H., Marshall, C. J.: EMBO J. *3*, 913 (1984)
313. Lax, I., Bar-Eli, M., Yarden, Y., Libermann, T. A., Schlessinger, J.: Proc. Natl. Acad. Sci. USA *81*, 5911 (1984)
314. Toh, H., Hayashida, H., Kikuno, R., Yasunaga, T., Miyata, T.: Nature *314*, 199 (1985)

315. Libermann, T. A., Nusbaum, H. R., Razon, N., Kris, R., Lax, I., Soreg, H., Whittle, N., Waterfield, M. D., Ullrich, A., Schlessinger, J.: Nature *313*, 144 (1985)
316. Crawford, L. V.: Int. Rev. Exp. Pathol. *25*, 1 (1983)
317. Reich, N. C., Levine, A. J.: Nature *308*, 199 (1984)
318. Eliyahu, D., Raz, A., Gruss, P., Givol, D., Oren, M.: Nature *312*, 646 (1984)
319. Parada, L. F., Land, H., Weinberg, R. A., Wolf, D., Rotter, V.: Nature *312*, 649 (1984)
320. Jenkins, J. R., Rudge, K., Currie, G. A.: Nature *312*, 651 (1984)
321. Kripp, B., Ferguson, B., Rosenberg, M., Westphal, H.: Proc. Natl. Acad. Sci. USA *81*, 6988 (1984)
322. Svensson, C., Akusjärvi, G.: EMBO J. *3*, 789 (1984)
323. Donoghue, D. J., Anderson, C., Hunter, T., Kaplan, P. L.: Nature *308*, 748 (1984)
324. Hurley, J. B., Simon, M. I., Teplow, D. B.: Science *226*, 860 (1984)
325. Keath, E. J., Caimi, P. G., Cole, M. D.: Cell *39*, 339 (1984)
326. Stewart, T. A., Pattengale, P. K., Leder, P.: Cell *38*, 627 (1984)
327. Pegoraro, L., Palumbo, A., Erikson, J., Falda, M., Giovanazzo, B., Emanuel, B. S., Rovera, G., Nowell, P. C., Croce, C. M.: Proc. Natl. Acad. Sci. USA *81*, 7166 (1984)
328. Tsujimoto, Y., Finger, L. R., Yunis, J., Nowell, P. C., Croce, C. M.: Science *226*, 1097 (1984)
329. Brown, N. A., Liu, C., Berenson, J. R., Garcia, C. R., Wang, R., Calame, K. L.: Proc. Natl. Acad. Sci. USA *82*, 556 (1985)
330. Flanagan, J. G., Rabbitts, T. H.: Nature *300*, 709 (1982)
331. Royer, H. D., Acuto, O., Fabbi, M., Tizard, R., Ramachandran, K., Smart, J. E., Reinherz, E. L.: Cell *39*, 261 (1984)
332. Toyonaga, B., Yanagi, Y., Suciu-Foca, N., Minden, M., Mak, T. W.: Nature *311*, 385 (1984)
333. Furley, A., Molgaard, H. V., Ford, A., Greaves, M. F.: ICSU Short Rep. *2*, 165 (1985)
334. Gale, R. P., Canaani, E.: Proc. Natl. Acad. Sci. USA *81*, 5648 (1984)
335. Collins, S. J., Kubonishi, I., Miyoshi, I., Groudine, M. T.: Science *225*, 72 (1984)
335a. Canaani, E., Steiner-Saltz, D., Aghai, E., Gale, R. P., Berrebi, A., Januszewicz, E.: Lancet i, 593 (1984)
336. Prywes, R., Foulkes, J. G., Rosenberg, N., Baltimore, D.: Cell *34*, 569 (1983)
337. Showe, L. C., Ballantine, M., Nishikura, K., Erikson, J., Kaji, H., Croce, C. M.: Mol. Cell. Biol. *5*, 501 (1985)
338. Nilsen, T. W., Maroney, P. A.: Mol. Cell. Biol. *4*, 2235 (1984)
339. Dani, C., Blanchard, J. M., Piechaczyk, M., El Sabouty, S., Marty, L., Jeanteur, P.: Proc. Natl. Acad. Sci. USA *81*, 7046 (1984)
340. Wiman, K. G., Clarkson, B., Hayday, A. C., Saito, H., Tonegawa, S., Hayward, W. S.: Proc. Natl. Acad. Sci. USA *81*, 6798 (1984)
341. Lachman, H. M., Skoultchi, A. I.: Nature *310*, 592 (1984)
342. McCormack, J. E., Pepe, V. H., Kent, R. B., Dean, M., Marshak-Rothstein, A., Sonenshein, G. E.: Proc. Natl. Acad. Sci. USA *81*, 5546 (1984)

Alteration of Blood Groups and Blood Group Precursors in Cancer

William J. Kuhns* **and F. James Primus**

Department of Pathology, University of North Carolina, Chapel Hill, N.C. 27514 and the Center for Molecular Medicine and Immunology, Newark, N.J. 07103, USA

Cancer-associated alterations of blood groups or their precursors occur most frequently in carcinoma and in leukemia. Antigens most often affected are those of the AB0(H), Ii, Lewis, MNT and P systems. The blood group changes in cancer may involve one system or multiple systems. When reported in leukemias, blood group deficiencies have usually been confined to A and H antigens, but in some cases, have been accompanied by other blood group system deficiences. Gene specified α-2-L-fucosyltransferase deficiency is a common precursor abnormality in untreated acute leukemia. The changes observed in adenocarcinomata and other solid tissue tumors are varied. Some tumor cell surfaces lack or are deficient in blood group determinants, whereas, gains of antigen have been reported in other tumors, at times inappropriately expressed relative to erythrocytic determinants. The probable causes relate to

1. production of simplified glycolipid or glycopeptide patterns in comparison with the normal tissue counterpart
2. production of unusual or atypical complex forms of glycoconjugate.

Current evidence suggests as a frequent cause deficient or aberrant synthesis of gene specified glycosyltransferases rather than excessive glycosidases. Synthesis of novel tumor related substrates may in turn provide favorable microenvironments for ubiquitous sugars such as sialic acid or other carbohydrates whose acceptor requirements may be less rigid than those carriers responsible for A, B or 0 specificities. In addition to changes in antigen status, malignancy can produce alterations in immune responsiveness as reflected by altered levels of circulating anti-A, anti-B, anti-T or other naturally occurring antibodies. Blood group-like autoagglutinins may be formed, in some cases of neoplastic disease and may cause active hemolysis. In many cancers, a heightened cell mediated immunity can be demonstrated using extracts of erythrocytic T antigen. The latter phenomenon has established a role for one blood group system in cancer diagnosis. Specific diagnosis of gastrointestinal cancers may be aided by blood or tissue tests utilizing monoclonal antibodies formed against Lewis-system-related cancer antigens. These and other monoclonal reagents are significant aids as applied to immunohistochemical studies of tumor tissue. Therapy of tumor using cytotoxic blood group antibodies or radiolabeled lectins has shown initial promise and may be expected to expand as hybridoma technology becomes more widespread.

* Present address: Hospital for Sick Children, Biochemistry Research, Toronto, Ontario, Canada

Progress in Clinical Biochemistry
and Medicine, Vol. 2

1 Introduction

Tumor transformation produces numerous antigenic alterations[1], particularly among the glycoconjugates, sugars linked to each other, to lipids and to proteins[2,3]. Many blood group antigens are identified as glycoconjugates; they include the AB0(H), MNT, Lewis, Ii and P antigens[4]. These determinants are particularly valuable since they can be understood in terms of their serology, chemical structure and biochemical genetics[4-7]. Their specific functions are not well understood; however, their position as terminal or penultimate sugars may assign them a special role as protective or informational molecules[8-10]. Ordinarily, these antigens, equally described on erythrocytes[410], persist unchanged throughout the lifetime of the individual, but they may be susceptible to alterations in the presence of certain hematopoietic diseases and solid tissue tumors, primarily carcinomas. A review of this topic was published by one of us in 1980[11]. New data has accumulated rapidly since then, particularly in the areas of structural analysis, biochemical genetics, and immunologic and tumor marker relationships. The purpose of this review is to provide an update of progress realized in these areas. Other reviews with emphasis on special areas are also available[12-15]. Current and prospective experimental studies should be of use in regard to the following questions:

1. are changes in blood group antigens unique concomitants of tumor transformation, or are they phenomena symptomatic of a more general event, such as altered molecular precursors, membrane domains or altered cell subpopulations?
2. are the changes essential to tumor initiation or tumor progression?
3. are they sporadic occurences in leukemias and carcinomas or are they predictable under conditions as yet poorly specified?
4. can the changes be reproduced in experimental cancers?
5. how can these cell surface alterations be of use in diagnosis or therapy?

2 Blood Groups and Blood Group Precursors – Overview

The AB0(H), MNT, Lewis, Ii and P systems account for five of more than fourteen major blood group systems which comprise more than one hundred antigens. There are, in addition, numerous private and public red cell antigens. They are observed by hemagglutination using operationally monospecific alloagglutinins[16]. Recently, highly specific monoclonal antibodies and affinity purified agglutinins have been prepared against several of the above antigens[17,18,402]. Such antibodies can be employed in standard hemagglutination tests[16], hemagglutination inhibition[19], immunoelectron microscopy[20], specific red cell adherence[21], enzyme linked immunoassay[22], or radioimmunoassay[23].

Specific terminal or penultimate sugar moeities account for the specificities displayed by these antigens[24]. Their attachment to carrier macromolecules is catalyzed by glycosyltransferase enzymes which transfer sugar from nucleotides[4,25]. The genes exist on separate chromosomes, but their products can result in sugar additions to the same macromolecule, which can be glycoprotein, glycolipid or oligosaccharide[4)].

A and B structures are characterized by terminal N-acetylgalactosamine (GalNac) or galactose (Gal), respectively, which are added to a common precursor possessing 0(L-fucose) specificity. The AB0 gene locus possesses two functional alleles, A and B, and a nonfunctional allele, 0(H); the alleles code for specific glycosyltransferases which catalyze sugar additions to H precursor[4, 6)]. Groups A and B are divided into subtypes, the major ones being A_1 and A_2, and A_1B and A_2B respectively. The enzymatic basis for these differences relates to qualitative[26)] and quantitative[27)] differences in the gene specified N-acetylgalactosaminyltransferases. A third independent locus, Se or secretor, codes for an α-2 fucosyltransferase which catalyzes the addition of terminal 1 → 2 linked L-fucose to penultimate Gal on a Gal-GlcNac disaccharide, thereby conferring blood group H specificity[412)].

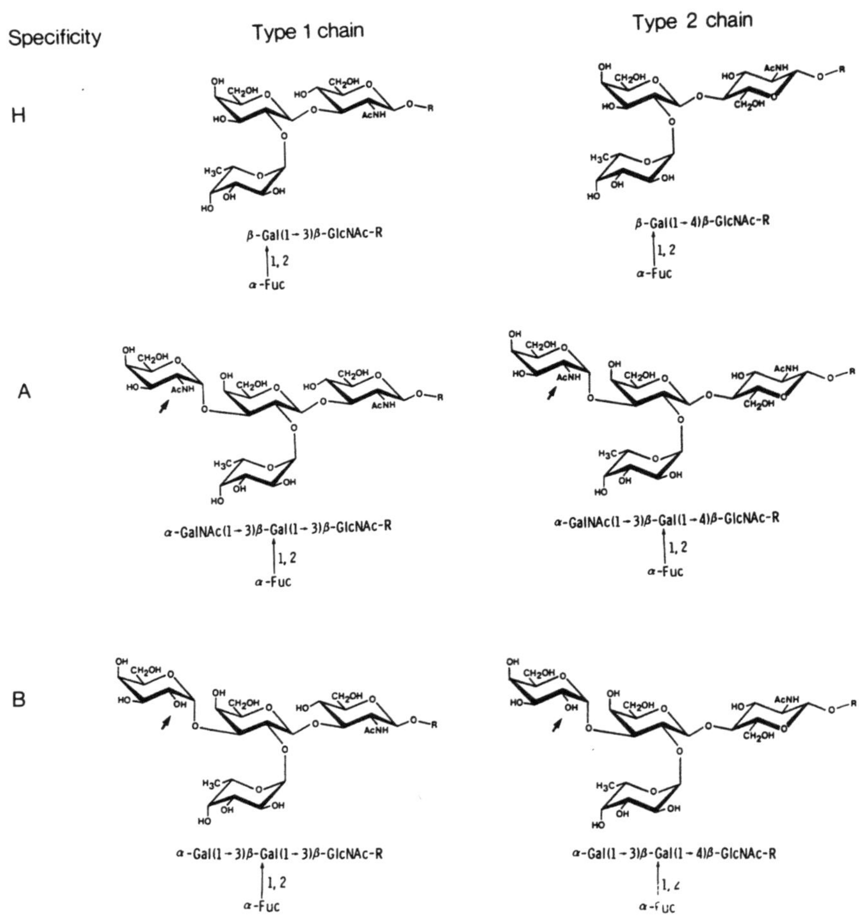

Fig. 1. Structures of H, A, and B determinants based on Type 1 and Type 2 carbohydrate chains. The arrows indicate the position in the terminal sugar ring where there is a difference between A and B structures. Abbreviations: Gal, galactose; GlcNAc, N-acetyl-D-glucosamine; GalNAc, N-acetyl-D-galactosamine; Fuc, L-fucose (From Watkins, W.: Adv. Hum. Genet. *10,* 1 (1980), reprinted with permission)

Specificity	Structure
H	β-Gal(1→3)GlcNAc ↑ 1,2 α-Fuc
Le^a	β-Gal(1→3)GlcNAc ↑ 1,4 α-Fuc
Le^b	β-Gal(1→3)GlcNAc ↑ 1,2 ↑ 1,4 α-Fuc α-Fuc

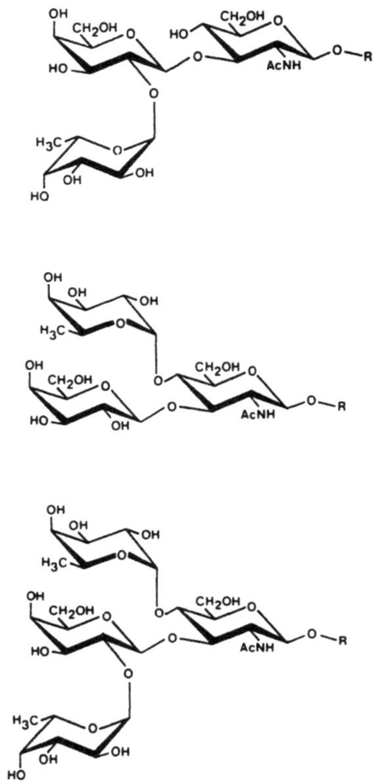

Fig. 2. Structures of Tpye 1 H, Le^a, and Le^b determinants. Abbreviations as in Fig. 1 (From Watkins, W.: Adv. Hum. Genet. *10,* 1 (1980), reprinted with permission)

Non-secretors (sese), who possess the Lewis gene which codes for α-1 → 4-fucosyltransferase form Le^a substance, L-fucose being attached in the 1 → 4 position on penultimate N-acetylglucosamine (GlcNac). Lewis positive secretors code for both of the above enzymes which produce a difucosyl compound on 1 → 3 linked Gal-GlcNac to form the gene interaction substance Le^b [4].

A and B antigens are formed on membrane structures or in secretions. The H and Le antigens are produced in secretions, including plasma where Lewis substance has been identified as a glycolipid[28], and can be adsorbed on to red cells[28]. The structural relationships of ABH and Lewis antigens are schematized in Fig. 1. The acceptor requirement for Lewis antigens is more stringent than for H, in that precursor Gal-GlcNac must be 1 → 3 linked, whereas for H, either 1 → 3 or 1 → 4 linkages provide a satisfactory substrate (Fig. 1 and 2)[4]. Recently, globoside has also been identified as a substrate for H sugar[411].

Quantitative analysis by radioimmunoassay of A and B antigens in saliva of ABH secretors and nonsecretors by LePendu et al.[23] has indicated that there is competition between A or B and Le gene-specified products. One B and two A specific radioimmunoassays detected A and B antigenic determinants in saliva of ABH non-secretors. The A or B antigens found in saliva of nonsecretors had lower efficiency and higher heterogeneity in the inhibition of the A or B assays, as compared to A or B antigens in saliva of ABH secretors. In the three assays, samples of non-secretor Lewis positive donors had less A or B than samples of non-secretor Lewis negative donors.

The disaccharides which furnish the penultimate sugars for ABH Le expression are terminal to linked oligosaccharides or to ceramide or protein linked oligosaccharides; prevalent chains among these carriers are linear lacto-series compounds, lacto-N-tetraosyl and lacto N-norhexaosyl[2, 6)] and branched repeating Gal-GlcNac compounds, lacto-N-isoctaosyl[2, 6)].

The Ii antigens can be regarded as differentiation products[29, 30)] and are not known to be derived from allelic genes as are the A and B antigens[16)]. Reactivity for i requires linear repeating Gal 1 → 4 GlcNac as in the lacto-N-norhexaosyl compound, whereas I requires in addition, branched lactosamine from 3 and 6 positions on penultimate Gal[6)]. The linear arrangement of i is prevalent on fetal membranes and on immature blood group substance[6, 31, 32)] whereas branched I substance characterizes adult blood group macromolecules[6)], probably the result of an activated branching enzyme[406)].

Fig. 3. Two novel gangliosides with branched carbohydrate chains (A, B) and the I-active human ganglioside (C) and I-active rabbit erythrocyte glycolipid (D) (From Hakomori, S.: Sem. Hemat. *18*, 39 (1981), reprinted with permission)

As in the case of competition between Se and Le gene products, fucosyl- and sialyl-transferases can compete for the above substrates as carriers for their respective sugars[2, 6)]. The expression on linear chains of i antigen will differ depending upon the terminal sugar, abrogation in the case of fucose[33)] and partial expression in the case of sialic acid[33, 34)]. Similar arrangements will affect I expression[35)].

Structures within the P system that can be built up from a common precursor include the P^k and P antigens. The precursor exists as glucosyl-ceramide (CMH) which serves as the substrate for lactosyl ceramide (CDH), the structure of p. The P^k compound (CTH) requires terminal α-Gal which is the carrier for terminal β-GalNac, the completed molecule being known as globoside (P antigen). The latter can function as an acceptor for Forssman antigen[5)] in the presence of an appropriate galactosaminyltransferase or for H antigen[411)] when penultimate β-galactose is present. The P_1 compound has been found on lacto-N-tetraosyl structures that are shared in common with ABH oligosaccharides, for example, paragloboside[5)].

M and N antigens are unique in that they are specified by sialic acid in non-covalent linkages with specific amino acids on sialopeptide molecules[7, 36)] where the oligosaccharide chains are comparatively short. Removal of sialic acid and exposure of penultimate β-galactose reveals T specificity which is normally cryptic[37–39)]. Removal of β-galactose exposes core N-GalNac with the emergence of Tn specificity[40, 41)]. Thus, T and Tn are MN precursors[36, 39)].

Tk and Tx are other cryptantigens with serological properties unique from T and Tn, each representing a product resulting from exposure to specific microbial glycosidases[42, 43)]. ABH, Le, Ii and P appear not to be expressed upon these sialoglycopeptides, although in the case of ABH, there seems to be no a priori reason why penultimate β-Gal could not serve as an acceptor for H specified L-fucose[4)].

In contrast to ABH terminal sugars which may appear on highly branched and complex oligosaccharides, M, N and T sugars exhibit a lesser degree of complexity and branching[7)].

Compounds onto which these blood groups and their precursor sugars may integrate themselves include, in the case of ABH, neutral glycosphingolipids (GSLs) and O-serine/threonine linked or N-asparagine linked glycoproteins[4, 6)]; Le and Ii, neutral and acidic GSLs and glycoproteins[4, 6)]; P system, neutral GSLs and glycoproteins[5)]. All of the above determinants can also be formed as constituents of complex carbohydrates such as mucins[4)]. MNT antigens are predominantly localized on sialo glycopeptides or glycophorins[7, 36)], but their occurrence on lipid has been reported[44)].

Organ distribution of some of the above blood groups or their precursors has been studied extensively, both by serological[45–48)] and biochemical methods[5–7, 49)]. By serological and histological methods, ABH, Le, Ii, P and MN have, apart from their location on erythrocytes, been localized on squamous, columnar and transitional epithelium, and the vascular endothelial lining of most if not all organs[50–54)]. Tissue blood group expression within organs is variable[55)]. Blood cells other than erythrocytes have exhibited blood group reactivities[56, 57)], although the adsorption of soluble blood group substance by lymphocytes and platelets may be proportionately greater than it is on red cells[58–60)]. Neutrophils have been stated by some workers to lack AB0 antigens on their surface[403)], and blood group H gene specified α-2-fucosyltransferase has not been demonstrated on these cells[404)]. Recent studies have suggested that lymphocyte subpopulations may be distinguished based upon differences in membrane glycolipids and blood group Ii antigens[61–63a]. Tissue distribution within kidney, pancreas, liver and placenta has been carefully studied for multiple antigens by immunofluorescence[50]. In kidney, for example, proximal tubules were negative for H, Le and Ii, whereas these antigens were strongly positive in collecting tubules. In liver, hepatocytes were negative, but biliary cells and Kupffer cells were positive for combinations of the above antigens[50]. Trophoblast tissue was found to contain small amounts of i antigen, but not other antigens such as ABH, I, Le and P[50].

Gastrointestinal mucosa contains large quantities of blood group substances in relation to other tissues. Of particular interest is the presence of both types 1 and 2 chains in contrast to red cells, which contain only type 2 structures[64]. Pancreas was found to contain only type 1 chains on fucolipid[49]. Blood group distribution diminishes distally in the intestine, such that ABH and Le^b is expressed in the proximal colon, but not in the rectosigmoid[47, 65, 66]. In contrast, Le^a structures have been demonstrated throughout the normal adult colon[67, 68].

ABH and other antigens may be cryptic in many cells and tissues[69, 70] due to masking by sialic acid or sialopeptides and are not demonstratable by conventional methods. Pretreatment of tissue with trypsin or sialidase followed by agglutination or staining has at times been successful in revealing these determinants. In certain tissues such as brain, type 2 Gal-GlcNac chains are common structures but ABH antigens are not expressed, probably because these substances occur as fucosylated gangliosides[71], with fucose attached 1–3 to penultimate GlcNac. Brain tissue also possesses a unique disaccharide, α-Gal 1 → 3 N GalNac[72], which is in contrast to β-Gal 1 → 3 N GalNac MN precursor found in most tissues studied. The factors accounting for normal blood group variability are probably complex. In this context, the role of sialic acid as a regulator of fine structure or as a rival for common substrates deserves further study.

Recently perfected methods of analysis coupled with the use of monoclonal antibodies against defined structures have quickened research in this area. As an example, erythrocyte receptors for mycoplasma microorganisms have now been characterized as sialylated Ii structures[73]. Monoclonal antibody prepared against a human epidermoid carcinoma cell line was found to precipitate the glycoprotein receptor for epidermal growth factor. Inhibition studies using the oligosaccharide lacto-N-fucopentacose I indicated the presence of type 1 blood group precursor chains[74].

Cell separation methods may assist in the assignment of specific precursor structures to specific cell types within tissues or organs. Lectin affinity chromatography may prove particularly useful in this regard. A recent technique makes use of a single affinity absorbent, hog gastric mucin group A + H substance covalently coupled to Sephadex or Sepharose, to which lectins of various specificities can bind, thus forming affinity probes[74a]. Other studies have demonstrated that fucose-binding lectin from Lotus tetragonolobus bound selectively to blood polymorphonuclear leukocytes, and that binding was prevented by L-fucose[74b].

3 Methods

Cancer associated blood group changes have been studied by three general approaches:

1. statistically by comparing the normal blood group incidence with that observed in cancer populations or families; the technique usually employed for such associations is standard red cell hemagglutination[16, 75];
2. serologically by mixed agglutination or specific red cell adherence[76, 77] or neutralization inhibition[78] in established malignancies in searches for alteration of red cell or tissue blood group reactivity or for mixed cell populations; and
3. chemically or histologically[2, 49, 79] by comparison of the structural or enzymatic status of blood group or their precursors in tumor vs normal adjacent tissue or tissue from unaffected individuals.

Although, chemical analysis is of profound importance in establishing fundamental changes of patterns in enzymes or in enzyme structure, its requirements for quantity impose restrictions on the types and numbers of carcinomas or other cancers available for analysis. This obstacle will probably disappear with the continued development of newer purification techniques, micromethods, and of monoclonal or affinity purified antibodies for use in isolation and staining by blot techniques[17, 18, 80–85]. Recombinant DNA technology has been utilized to clone genes for bacterial glycosylltransferases and these methods should prove important in future studies of precursor synthesis utilizing appropriate vectors and cell mutants.

Tissue section analysis may be carried out separately or in parallel with biochemical methods using antibodies as described above which have been appropriately labelled. Among the most popular labels are fluorescein or rhodamine employing ultraviolet microscopy, or horseradish peroxidase utilizing light microscopy[86, 87]. A variety of bridging techniques are available, which serve to enhance specific tissue reactions[22, 392]. A method frequently employed in immunodiagnosis is specific red cell adherence[51] where specific antibody acts as a bridge between indicator red cells and surface epithelium. Immunohistological methods are of special value since they are extremely sensitive, they require only small amounts of tissue, large numbers of specimens can be examined simultaneously, and antigen distribution can be related to histopathologic changes[87]. The use of more than one sugar specific antiserum or lectin or an examination of adjacent sections by diverse antibody reagents may aid in simultaneous studies of blood group terminal sugars and their acceptors[88, 89]. Processing techniques utilized for histopathology may prove injurious to certain antigens and thereby restrict interpretations of tissue studied retrospectively[45, 46]. Alcohol soluble blood group substance associated with lipid may be removed in the process of paraffin embedding, but water soluble materials of epithelial mucins will be preserved[45, 46]. Thus, prospective studies may benefit from comparisons of stained tissue which has been frozen with material fixed by conventional methods.

4 Blood Groups in Cancer – Serologic Studies

A. Blood Group Associations with Cancer

Blood groups are related to disease susceptibility, but the exact mechanism of this relationship is unclear. It may be that force of natural selection is acting through disease to affect the various blood group frequences[90]. For many years, deficiencies in the techniques or reagents employed in blood grouping made it difficult to evaluate numberous claims which followed Landsteiner's discovery of AB0 antigens. By the 1950s, most reliable investigator's were working with standard typing methods, using high-titered reagent antisera. During this time, and the decade that followed, several important associations were established between blood groups and disease. Those associations involving diseases other than cancer will not be mentioned here, but are summarized by various authors[91–99]. The first well-documented claim was established between erythrocytic blood group A and carcinoma of the stomach[100]. The data were based upon comparisons between individuals with this disease and a normal population of equivalent size, and indicated that if the risk that a person of blood group 0 or group B will develop carcinoma of the stomach is taken as 1, then the risk for a group A individual is 1.2. Observations which now total over 63,000 indicate that this, and that for carcinoma of the ovary, are the most firmly established associations[91].

Mourant et al.[91] have indicated that nearly all carcinomas are in some degree associated with blood group A, and that in most cases, the association is statistically significant. Associations with secretor A also appears to exist, but are based upon relatively small numbers. In contrast, associations of group B with carcinoma have usually not been statistically important.

Laboratory support for these findings has been difficult to obtain, but it is interesting that A-like Forssman glycolipids have for many years been thought of as tumor-associated agents[101–103]. Hakomori et al.[104] recently prepared Forssman-like chemical extracts from stomach and colon carcinomas which were similar to A antigen. Levine (1978) has indicated that correlations exist between carcinomas and low quantities of circulating anti-Forssman hemolysins which may exist as circulating immune complexes. Other workers have found no correlation between the latter and anti-Forssman titers[106]. It remains to be determined whether anti-Forssman antibody could perform tumor surveillance more effectively when A-like tumor cells arise inappropriately in 0 and B persons, as compared with their occurrence in A individuals.

Leukemias have occasionally been accompanied by weakening of erythrocytic blood group antigens (discussed below), whereas in carcinomas, only the tumorous tissue appears to be altered. The results of conventional blood typing have shown that blood group distribution among leukemic patients is not significantly different from normal, although many leukemic patients with altered blood group expression have been described since 1957. Thus, among 590 patients investigated by Gunz (personal communication), there were no deviations from the normal pattern in the ABH, Rh, MNSs, Kell, Lewis, and Lutheran systems.

There appears to be a strong selective effect of the AB0 groups on the development of choriocarcinoma. In comparison with women who are of group 0, the risk of choriocarcinoma for group A women is less if the husband is also of group A, but is greatly increased when he is of group 0[108, 109]. The risk does not seem to be related to the AB0 group of the fetus, since choriocarcinoma-associated children born from A × 0 matings have been equally divided among group 0 and group A. The effects of blood groups on the development of choriocarcinoma would therefore appear to be influenced by genetic factors reflected in the AB0 group of the husband, but not in the AB0 group of the conceptus, raising the question of whether this effect occurs at or before fertilization.

In an interesting family study of more than twelve hundred persons, in which seven cases of Hodgkins disease were documented, there were twice the expected number of Rh negative individuals[110]. This raises the question whether certain Rh haplotypes predispose to Hodgkins disease via gene linkage, or whether alternative Rh haplotypes confer disease resistance.

B. *Deficient Blood Group Expression in Leukemia*

Blood group red cell antigen modifications resulting from a malignancy were first observed in acute leukemia[111–118]; all of these studies described weakened expression of A antigen in the course of myeloblastic leukemia. Additional case reports of a similar nature established group A as the most frequently affected of the blood groups. Although most of the published reports concern acute myeloid leukemia (AML), other cytologic forms of acute leukemia have been associated with the same red cell defects[119]. Secretor activity appears to be present in many cases with weakened erythrocytic antigens[111, 114, 115, 118, 120] and absent in others[112, 113, 116], but in some cases has not actually been determined.

After group A, Group H appears to be the next most frequently involved antigen[107, 119, 121] and a few reports have described loss of B antigen[122–124], I-i alteration[125–128] and modifications of Le^a I[117], P[129], MN[130], and Rh[130–132]. Abnormalities of γ-globulin groups have occasionally been described[133, 134] indicating the involvement of lymphocytes themselves. In three cases of acute leukemia, multiple antigen loss was described involving A and Le^b antigens, and in two cases i antigen[135]. Other multiple antigen losses have been described[130–131]. With very few exceptions[119, 136], the reported cases have described a weakening rather than an increase in strength of the blood group antigens.

Salmon[119, 137] who has studied this subject extensively, has classified weak (group A) reactors into three classes, according to the quantity of A-reacting cells detectable in a mixed normal A-weak A, mixed A + non-A or serologically nonreactive population. Presumably, these findings reflect the presence of one population derived from normal precursor cells, and a second population derived from the erythroid progeny of leukemic cell precursors; in such circumstances, the proportion of normal to abnormal cells can range from large to small. The leukemic cells themselves may demonstrate altered expression of blood groups[138], although there is some evidence that such alterations may be the consequence of masking by sialic acid[139]. An additional category described by Salmon and others[121] is comprised of cases which demonstrate imperfect production of H antigen. Such cases are analogous to the Bombay group, in which an amorphic gene results in a lack of H expression owing to the absence of an α-2-L-fucosyltransferase[140]. The result is an inability to express A or B, even in the presence of the corresponding transferase enzyme. Some cases of weakened group A expression could be accounted for on this basis, i.e., incompleted acceptor molecules on abnormal cells would be incapable of functioning as normal acceptors for A or B sugars. Most acute leukemic patients appear to be deficient in circulating α-2-L-fucosyltransferase[141], and this may relate in turn to abnormalities of blood group acceptor molecules, although phenotypic expression on circulating erythrocytes can be demonstrated in many of these cases.

The possibility exists that in some instances steric modification of cell surface macromolecules associated with altered morphology may be a cause of blood group loss, rather than enzyme abnormalities[142, 143], such that precursor functionality could be impaired in the presence of normal enzyme.

An excess of i antigen has been reported in acute leukemia, a finding that could demonstrate a relationship of this marker to marrow transit time as reported for thalassemia major and paroxysmal nocturnal hemoglobulinuria (PNH) by Giblett and Crookston[393], but could also be related to the differentiated status of leukemic cells or their precursors[144].

Kannagi et al.[143–144] studied the differentiation of mouse leukemia cells and found sequential shifts of glycolipid species from short ganglio-series synthesis and linear chains to elongated branched lacto-series types which could be determined by i to I antigen transitions. The major glycolipids in mature human blood cells are globo- and lacto-series, whereas leukemic cells tend to display i substance relating to ganglio-series glycolipid[2].

The i antigen on lymphocytes, in contrast to its counterpart on mature adult erythrocytes, is easily detectable in normal adults. However, in chronic lymphocytic leukemia (CLL) i antigen is markedly decreased compared to the normal, a finding potentially useful when distinguishing CLL from other forms of lymphocytosis[145]. Tests with anti-i may also be helpful in identifying the lymphoid or myeloid origin of blasts in undifferentiated leukemia since lymphoid blast cells possess greater amounts of i antigen than do myeloid blast cells[128].

Bird et al.[146] studied a patient with acute myelocytic anemia in which a mixed population of erythrocytes was caused by Tn polyagglutination, a rare erythrocytic abnormality which may occur in otherwise unaffected persons. The case was characterized by two erythrocyte populations, Tn and non-Tn. The former disappeared during the course of the patient's disease, probably due to removal of the corresponding clone of stem cells in association with cytotoxic drug therapy. Further reports have been described associating Tn polyagglutinability with leukemia[146, 147]. The polyclonal nature of leukemia is suggested from observations that the various markers can occur on cell populations independently of one another[119].

In some cases, red cell abnormalities observed at presentation were noted to revert to normal during remission, but again changed following relapse. In one exceptional instance[148] a modification of a blood group was reversed to normal as the patient experienced a terminal relapse and the number of abnormal white cells rapidly increased.

Salmon[119] has demonstrated that blood group losses, deficiencies or modifications of markers other than AB0 are not restricted to the leukemias but appear as well in refractory anemias and other preleukemic states, and on occasion in elderly and normal subjects. Such observations may correlate in part with those of Parker et al.[407] who noted abnormalities of glycoprotein domains on erythrocytic extracts derived from PHN patients, many of whom ultimately develop leukemia. Quantitative agglutination studies of refactory anemia by Salmon et al. indicated modifications of A, B, H, I, and i antigens in 10 of 11 cases, which suggested a relationship between this disease and leukemia. Kahn et al.[149] studied a case of erythroleukemia in a patient who was originally blood group A_1. In the course of the disease, 50% of the A antigen was lost from his erythrocytes. Separation of the abnormal cells showed changes in several enzymes, including adenylate kinase, which interestingly, is closely linked to the AB0 genetic locus[150].

The incidence of blood group abnormalities in leukemia has been evaluated using quantitative agglutination techniques. In two series of cases, H antigen deficiencies were reported in 9 of 81 (11%) cases of malignant hemopathies, and I-antigen modifications in 24 of 105 (23%) cases[127, 137]. Ayres et al.[131] demonstrated abnormalities in 9 of 51 leukemic patients (18%), the more common deficits being related to the H and I antigens. Transfusion in cases of severely anemic leukemia patients was cited as a potential problem in evaluating the strength of blood group antigen, and it is obvious that this would dilute any leukemogenic effects.

Transfusion of normal A cells into a leukemic patient with weakened A expression was utilized by Salmon et al.[151] to determine whether an intrinsic factor such as an A specific glycosidase was capable of modifying normal group A cells. Tests of separated normal transfused cells indicated that they had retained their original antigenic strength.

In a study by Majsky[152] the percentage of red cells not agglutinated with anti-H was studied in group 0 patients with acute leukemia. Abnormal values (designated as more than 5% of unagglutinated erythrocytes) were found in 27% of the cases. The cases included decreases in agglutinability of all corpuscles, as well as mixed populations containing cells with normal H antigen and cells with decreased amounts of H antigen. Salmon and Cartron[153] found that 20% of leukemic patients had lower than normal anti-H scores using anti-H reagents. Chemotherapy in leukemia has been cited by one group as the major cause of A-antigen changes or the appearance of dual populations[154], but the predominating opinion appears to be that remission, in most instances thought to be initiated by treatment is accompanied by a reversion toward normal in individuals with previously weakened blood groups[107, 119]. Presumably, the changes reflect shifts in a mixed normal/leukemic cell population.

An important aspect of changes in blood group phenotypes or their precursors in acute leukemia and blastic crisis is that they specify abnormalities in genetic function. The following evidence is in support:

a) deficiency in the gene specified transferases can be observed[155]
b) Gm immunoglobulin modifications sometimes correspond to one of the haplotypes of the stem cell and since allotypes specify different Ig molecules, modifications of different Ig markers must be produced by stem cells prior to functional differentiation into cellular progeny[156].

Chromosomal translocation of chromosomes known to carry blood group genes occur in leukemia, more in chronic myeloid than acute leukemia, although aberrant AB0 expression seems to occur more frequently in the latter condition[157]. Lines of leukemic cells now available for culture studies raise the future possibility of exploring these relationships in detail[144].

C. Carcinoma Related Blood Group Alterations

The demonstration of blood group activity in normal secretions and on normal tissues other than erythrocytes[47, 158–162] was followed by studies of secretions and tissue-associated antigens in a variety of diseases. It was found that carcinomas were frequently associated with impaired blood group expression at, or in proximity to, the sites of neoplastic transformation, but not on erythrocytes of these patients. Early studies of defective expression of A and B antigen losses in glycolipids and glycoproteins of human adenocarcinomas have been described by Ohuti[163], Masamune et al.[164], Kay and Wallace[165], Kawasaki[166], Davidsohn et al.[167, 168], Dabelsteen and Fulling[169], and Hakomori et al.[170]. The conclusions were based on differences from the normal, making use of tissue extracts and immunochemical methods[163, 164, 166, 170], or using techniques of agglutination or mixed hemagglutination on cell suspensions[165, 285], and immunofluorescence[167, 168] or specific red cell adherence in fixed tissue sections[167, 168].

Kay[171] in studies of epithelial cells derived from the normal urinary tract, found that the cells reacted in a mixed agglutination test, in conformity with the individual's erythrocytic AB0 group. However, abnormal epithelial cells derived from scrapings of 25 carcinomatous bladders, ureters, and a renal pelvis reacted poorly or not at all in many cases. Five tumor cell suspensions failed to absorb the blood group antibody that corresponded to the expected blood type, and this was accompanied by loss of agglutinability of these cells in comparison with their normal counterparts. An approximate inverse relationship was demonstrated between the degree of agglutination and the malignancy of the tumors studied as judged by microscopy and the extent of tumor infiltration or tumor metastasis.

Glynn and Holborow[52] utilized specific immunofluorescence to study the distribution of blood group substances in tissues. They reported specific immunofluorescent staining of blood group substances in cancer cells of the stomach. The specific reactions were lost after treatment of the sections with ethanol. This was interpreted to mean that the cancer cells more closely resembled parietal than superficial cells, because the parietal cells gave a positive fluorescent reaction in frozen sections. Eklund et al., using a indirect fluorescence technique, observed staining in 5 of 6 cases of gastric carcinoma with various degrees of differentiation. These studies were performed on frozen sections in order to preserve alcohol-soluble blood group substances[172].

Losses of AB0 antigens were later reported in urogenital malignancy[168, 173]. Confirmatory studies have shown AB0 loss in tumor-transformed bladder tissue[174]. Comparisons made between A and B antigens in normal and carcinomatous cervix have generally shown tumor-associated antigen reduction or loss[175].

Benign and malignant breast lesions studied for A and B antigens by specific red cell adherence showed consistent antigen deletion in intraductal carcinoma as well as in fibrocystic disease, which suggested to the authors a possible link between these conditions[176]. On the other hand, deletion of MN antigens and exposure of T appears to distinguish between these conditions since it occurs only in breast carcinoma and not in benign breast disease[177].

Davidsohn et al.[167] utilized immunofluorescent techniques to investigate the relationship between the dedifferentiation of carcinomas and the presence of water-soluble blood-group-specific substances. Tissues of the stomach, small intestine, ascending and transverse colon, and proximal portion of the descending colon in 81 adenomatous polyps and adenocarcinomas were compared with controls, utilizing the intensity of fluorescence as a criterion. In the absence of cancer, blood group substances were abundant in the epithelial cells and in secreted mucus. In cancer, the antigenic substances were absent in secreted mucus, presumably an indicator of dedifferentiation. Since these were retrospective studies, the secretor status of the individuals from whom the tissue was derived was unknown.

Davidsohn et al.[173, 178] later introduced specific red cell adherence, a modification of the Coombs mixed agglutination technique, to study a variety of carcinomas and normal tissues. The test appeared to be sensitive and specific, and unaffected by the age of the paraffin-embedded tissues and hematoxylin-eosin-stained slides. This permitted the authors to carry out retrospective studies on embedded tissues which had been processed as long as 35 years previously. The loss of antigens increased progressively from carcinoma *in situ* to anaplastic, invasive, and metastatic carcinoma, and was interpreted as evidence of immunologic dedifferentiation with anaplasia. Three-hundred-and-fifty primary carcinomas of these organs were studied. With few exceptions, the loss of the isoantigens preceded the formation of distant metastases. On this basis, the authors believed that the test was valuable in the diagnosis of early carcinoma in tissues that normally contain ABH antigen, and in estimating prognosis. Sheahan et al.[179] utilized Davidsohn's technique and found deletion of epithelial ABH antigen in primary gastric neoplasms and in metastatic cancer.

Subsequent workers have provided ample confirmation of blood group loss in gastrointestional carcinoma, as well as carcinoma in other locations[80, 180–181]. Terminal A or B antigen loss has frequently been accompanied by an accumulation of H antigen precursor[88, 182], Lewis antigen[170], or I antigen[183]. In recent studies, Dabelsteen et al.[89], who previously reported A and B losses in oral malignancies, used monoclonal antibodies against H antigen and its precursor N-acetylactosamine and showed that A, B losses were accompanied by accumulations of precursors.

Cancer-related blood group loss has been but one manifestation of malignancy. Gains of such antigens and at times their inappropriate expression in carcinoma have frequently been reported[102, 184–188]. This has been particularly true in areas such as adult distal colon, where A and B antigens are normally not expressed[185–187]. Denk et al.[185] have shown a re-expression of serologic activity in many colonic cancers. Cooper and Haesler[186] found that 8 of 17 adenocarcinomas of the distal colon possessed blood group substance, all in A patients. However, no colonic blood group antigen was found in group B patients.

Other cases of cancer have been described in which acquisitions of A antigen occurred in non-A patients[187–188]. Paul et al.[189] reported in two glycopeptide fractions from an ovarian cystadenocarcinoma and an appendix tumor, from patients with blood groups B and 0, respectively. The glycopeptides, in addition to demonstrating compatible B and H activities, also demonstrated incompatible A activity. Runge and Pour[190] used a chemical carcinogen to induce pancreatic tumors in hamsters, and found that mucins produced by the tumor possessed A antigenic activity, although the erythrocytic group of these animals were 0-like. Yokota et al.[191] studied a hepatocarcinoma from a group 0 individual and have isolated two short chain glycolipid fractions, one A-like and the other containing strong Forssman antigen activity. Piller et al.[289] reported on an increase of blood group A and concomitant loss of Sda activity in mucus derived from a human neoplastic colon.

Studies by Yonezawa et al.[387] on the binding of *Ulex europeus* (UEA 1 or anti-H) lectin in distal colon sections are of interest. Binding of UEA could be demonstrated to normal proximal colon but not to normal distal colon. However, colonic mucosae became UEA 1 positive in adenomas and adenocarcinomas of the distal colon. In four cases of rectosigmoid polyposis, non tumor mucosa bound UEA 1 lectin[388].

The underlying basis for such discrepancies may be complex. It is comparatively straightforward to specify absent or deficient glycosyltransferases in cases of blood group loss[153, 154] although there are alternative explanations[142, 192]. In cases of A-antigen gain or inappropriate expression, cross reactions of tissue Forssman antigen with anti-A must be considered. It is of interest that in 32 cases of gastric cancer (including 15 group 0 cases) studied by the immunoperoxidase technique using monoclonal anti-A or anti-B, there was no evidence of inappropriate red cell antigen expression[181]. In 6 of 17 A or AB patients, A antigens were completely lost, although loss could not be related to degree of tumor differentiation. The authors inferred that inappropriate A expression could be recognized by conventional polyclonal anti-A antisera but not by their monoclonal reagents, and they are investigating this possibility.

Apart from such reasons, other biochemical causes could result in blood group gain. Alternate or novel precursors could compete successfully for limited blood group enzymes and sugars. In the case of A or A-like expression, several forms of substrate are known which could accept N-GalNac with A-like reactivity[2, 4, 9, 153]. There is also evidence *in vitro* that A and B-gene specified glycosyltransferases are less rigid in their donor substrate requirements than was previously supposed. Thus, conditions can be manipulated so that B-gene specified enzyme catalyzes the transfer of N-GalNac from UDP-GalNac to synthesize A determinants from H active structures[193]. It is not known whether this happens *in vivo* in cancers or any other disease process.

Zweibaum et al.[194] have reported gains of A and B determinants in cancer of the distal colon and have suggested that they may be markers for enzymes such as disaccharides. The latter, which can possess ABH determinants, and are normally located in small intestinal mucosa[195] may occur in ectopic locations in patients with colon cancer. This interpretation suggests that re-expression of ABH in distal colon may be associated with the presence of novel substrates associated with tumor growth.

The frequent variable nature of ABH behavior in gastrointestinal cancers precludes their use in diagnosis or prognosis. However, their usefulness in urinary bladder carcinoma has been demonstrated. Decenzo et al.[196] originally reported a relationship between ABH and the biological behavior of this tumor. The observations were later confirmed, and they have indeed provided urologists with a helpful diagnostic and predictive tool. Studies of bladder tissue by specific red cell adherence[12] and by immunohistology show that tumor invasiveness or non-invasiveness is reciprocally related to ABH blood group expression, thus permitting discrimination between superficial carcinomas at low risk for invasion from highly invasive tumors which justify aggressive therapy. Studies of blood groups in upper urinary tract cancers using similar methods do not appear to possess a similar degree of diagnostic or prognostic value[197].

Cell surface AB0 antigens have not been demonstrated in prostatic carcinoma tissues in contrast to normal or hyperplastic prostatic tissue, although there were no correlations with histologic grade[401]. Since the need to distinguish hyperplasia from carcinoma may at times prove useful on clinical grounds, the presence or absence of blood group phenotypes may serve as an important indicator[401].

The blood and tissue occurrence of Lewis-related phenotypes has been studied by monoclonal antibodies in patients with gastric, pancreatic and colon carcinomas[198, 199]. The antibodies react with gastrointestinal cancer antigens (GICA) which depending upon the individuals Lewis phenotype, forms a sialylated monofucosyl or difucosyl ceramide linked structure[1, 200, 201]. Sera of 64% of patients with colorectal cancer, 92% of patients with pancreatic cancer, and 72% of patients with gastric cancer inhibited binding of antibody to GICA-containing cancer cells. In contrast, sera from malignancies other than gastrointestinal or from normal subjects inhibited in only 8% and 2% of cases, respectively. In instances where GICA exists, Le antigen is masked by sialic acid and is presumably undetectable using anti-Lewis reagents. Our experiences with Lewis antigens in colon cancers have shown that Le^a normally present in proximal and distal colon is frequently deleted, whereas Le^b with normal distribution resembling ABH, may be gained or lost depending upon localization, and is thus similar to ABH in this respect[202].

Further studies relating cancer to the Lewis system have been reported by Hakomori et al.[203]. The author isolated glycolipids from adenocarcinomas which possessed unique terminal sugar linkages characterized by fucose attachment in the $1 \rightarrow 3$ position to penultimate N-GlcNac of type II chains (Le^x determinant). This structural unit resembled the stage specific embryonic antigen found in fetal tissue but not found in normal adult tissue[408]. A difucosylated structure (possessing an added fucose attached $1 \rightarrow 2$ to penultimate galactose) termed Le^y was also found as a unique concomitant of adenocarcinoma tissue. A monoclonal antibody (AH-6) prepared against a gastric cancer cell line was found to react with a series of glycolipids with the determinant and found in tumors and 0 erythrocytes[204]. Presumably the genes which specify glycosyltransferases catalyzing additions of such sugars to acceptors may be activated by certain transforming agents.

The possible role of the P system in cancer was raised by Levine et al. who studied the gastric cancer of a patient with the rare blood group phenotype pp (Tj^a or $P_1P_2P^k$ negative)[205, 206]. The lyophilized tumor was found to absorb anti $P_1P_2P^k$ antibodies, thus indicating that the tumor tissue was P positive in contrast to the patients erythrocytic group. Normal stomach from the patient was not available to confirm the absence of the antigen, but other specimens of normal gastric mucosa and three additional gastric carcinomas tested absorbed anti-Tja. Breiner et al.[207]

since then have studied normal gastric tissue from pp individuals and have demonstrated P-like antigens. This suggests that normal/tumor tissue distinctions may be quantitative rather than qualitative in nature. It is becoming apparent that the tissue distribution of the different P system antigens are normally variable, a circumstance requiring appropriate controls when analyzing cancer tissue. One aspect of the study by Kannagi et al., namely, the findings of P activity in tissue from this pp patient is difficult to reconcile with earlier work by Fellous et al.[208] who detected no evidence of P gene in pp fibroblasts.

The long term post-operative survival in Levine's patient raised the possibility that her longevity was due to a brisk immune response to a small transfusion of Tja positive red cells. It was postulated that any remaining tumor cells could have been eliminated by cytotoxic anti-Tja antibodies[209]. Kabat et al.[210] have reported a similar interesting case of an inoperable carcinoma of the lung with good survival in which the patient formed monoclonal antibody against an unusual, apparently tumor related blood group precursor with the determinant structure D-Galβ 1 → 3 D-GlcNacβ 1 → 3 D-Gal[211].

T antigen masked as cryptantigen is part of most normal human cell surfaces where it is generally shielded by sialic acid[37, 212, 213]. In contast, unmasked T structures can be demonstrated in many carcinomas[213] and in melanoma[214]. Springer et al.[215] have found that approximately 90% of 127 human adeno-small cell and squamous cell carcinoma membranes readily absorbed anti-T antibodies. Of those which did not, greater than 90% possessed Tn specificity, and usually were less differentiated than T containing tumors. Metastases were clonal in nature, since they had Tn and T if the primary lesions were also Tn and T positive. Anaplastic carcinomas usually possessed Tn or Tn in excess of T[216]. Similar results were obtained with lesions of the breast, lung, pancreas, gastrointestional and urinary tracts, larynx and orpharynx, neck and salivary gland. Healthy or non-carcinoma tissue only rarely possessed reactive T or Tn antigen.

In addition to their demonstration by absorption, these antigens have been revealed by a variety of serological and chemical methods, more recently employing monoclonal antibodies[217]. Peanut lectin, an anti-T-like reagent which cross reacts with T antigen was studied in 33 rectosigmoid cancers and 15 controls[218]. Formalin fixed sections stained by immunofluorescence demonstrated increased synthesis of T antigen in cancers as compared with controls. The behavior of this MN precursor therefore resembled that of A and B precursors (described above).

Results similar to these have been obtained by other investigators[219]. The results of immunohistochemistry in urinary bladder cancers have established important correlations between T antigen or cryptic T antigen and tumor invasiveness[220]. Similar studies have been carried out in benign and malignant breast tumors[221].

There has been a recent study of cancers using erythrocyte T antigen as a marker[222]. The rat monoclonal antibody for this agglutination test was prepared against desialated human red cells. It yielded positive results on red cells from patients with a variety of cancers, including leukemias, but mostly negative results on cells from healthy or non-cancerous individuals. Since microbial sialidases are capable of rendering erythrocytes polyagglutinable[9], further studies are required to rule out alternate possibilities.

D. Other Malignancies

Normal or neoplastic tissues in which blood group antigens have not been demonstrated include muscle, connective tissue, central nervous system parenchyma, testis, liver parenchyma and adipose tissue[12, 16, 45, 46, 47, 52)]. Blood group precursor-like material in human neuroblastoma cells was demonstrated by Samter and Glick[223)]. Membrane glycopeptides were found to possess fucosyl residues on their oligosaccharide antennae. By specific fucosidase digestion or by nuclear magnetic resonance the fucose linkages were located on $1 \rightarrow 3$ and $1 \rightarrow 4$ positions on penultimate GlcNac and therefore resembled the composition of blood group substance. Non-tumor brain glycopeptides have been shown to contain relatively high amounts of terminal non-substituted galactose and N-acetylglucosamine, suggesting that the peripheral branches are often incomplete and therefore similar to blood group precursor substance[224)].

Blood group antigens ABH have been evaluated in vascular tumors using the immunoperoxidase technique[225)]. No differences were noted between benign hemangioendotheliomas from children or adults; neither tumor exhibited antigen in intercapillary cells. Cases of Kaposi's sarcoma did not show antigens in spindle cells or capillaries, but in medium sized vessels, variable preservation or loss of blood group isoantigens was found. Odontogenic tumors or ameloblastomas have been examined for ABH antigens by immunofluorescence. There was a lack of A and B in fifteen patients in contrast to normal oral mucosa in which these antigens were present. All but one of these patients demonstrated H antigen in normal and tumor mucosa. Weakening of A and I antigens has been described in Hodgkin's disease, but this could possibly be related to the sequelae of radiochemotherapy[22)]. T and Tn have not so far been found in sarcomas, benign tumors or malignant tumors of the central nervous system[217)].

5 Immunological Changes Resulting from Altered Blood Groups in Cancer

Cancer-induced changes in antigens may enable some of them to behave as autoimmunogens. If immune responsiveness is not depressed, as is sometimes the case in malignancy[227, 228)], antigen changes induced by malignant transformation may be followed by humoral antibody formation and circulating immune complexes, or the occurrence of cell mediated immunity[229)]. The antibodies in our area of interest may be polyclonal or monoclonal[230)] and related to specific blood group structures[231, 232)] or they may possess other dominant specificities, but cross react with the latter[233)].

A. Humoral Antibodies

Alterations in the ability to form blood group antibodies in malignancy have been reported; hypogammaglobulinemia with reduced levels of anti-A and anti-B may be associated with malignant blood disorders[9]. Hamilton-Fairly and Akers[234] demonstrated marked reductions in anti-A and anti-B isoagglutinin titers in all forms of reticulosis, and in malignant diseases of lymphocytes and plasma cells. In chronic lymphatic leukemia (CLL) there was a definite relationship between gamma globulin level and the anti-B titer, and in group 0 patients between the anti-A and anti-B titers. In the other reticuloses, the anti-A titers were similar to the controls, but anti-B titers appeared to be slightly but significantly reduced.

Immunization with blood group A and B substances showed that the ability to form both anti-A and anti-B hemolysins was grossly impaired in malignant diseases in lymphocytes and plasma cells, in contrast to the other reticuloses, in which this ability was normal. Patients with Hodgkin's disease did not form hemolysins as well as those with myeloproliferative syndromes, but the difference was not significant.

Untreated CLL patients with hypogammaglobulinemia have been studied and were shown to be markedly defective in the generation of immunoglobulin secreting cells. Normal or increased T-cell helper activity was found without excessive suppressor activity, suggesting that B cells possessed an intrinsic defect[235]. However, in acute T-lymphoblastic leukemia with hypogammaglobulinemia, cells possessed a unique surface phenotype and were capable of suppressing Ig production of normal cells in co-culture experiments[266].

The results suggest the existence of two immunologic categories in patients with reticuloses;

1. those with malignant diseases of lymphocytes and plasma cells, in whom antibody production is greatly impaired in both primary immunization and reimmunization, and who frequently have low isoagglutinin levels and hypogammaglobulinemia[236–239].
2. those with other reticuloses, who show impaired antibody formation on primary immunization, but not on reimmunization, and who rarely have low isoagglutinin titers or hypogammaglobulinemia. This finding may be symptomatic of immune deficiency in the tumors of lymphoreticular origin, possibly as a caustive factor, since elimination of cancer-inducing agents may be reduced when antibody responses are impaired.

Deficiencies or loss of AB0 isoagglutinins in leukemia are uncommon. However, these isoagglutinins are found to be low in many chronic leukemia cases[238], perhaps because they have a lymphoreticular origin[119]. Qualitative changes of anti-B have been found in the course of AML. The techniques were based on the observations of Filitti-Wurmser et al.[240–242], who carried out studies on the relationship of AB0 blood groups to the thermodynamic properties of anti-A and anti-B isohemagglutinins. This data was used to study the effects of antigen modification in leukemia on the properties of anti-B[243]. Measurements carried out for 2 years in a 5-year old nonleukemic group A_1 child with erthroblastopenia who received alternating transfusions of washed group A_1 or 0 cells, showed reversible changes of anti-B affinities, i.e., increases following A_1 cell infusions, and decreases subsequent to group 0 infusions. Alternating anti-B affinities in leukemic patients with modified A groups were thus explained as the consequence of chemical changes of A antigen caused by the leukemic process. Presumably, A-bearing molecules which possess structures closely related to the B antigen, and which differ by only one terminal nonreducing sugar, can affect the affinities of anti-B, but abnormal changes in these anti-B properties are observed when patients' antigens become 0-like. A decrease of H antigen in chronic myelocytic leukemia (CML) with undetectable anti-A,B was believed to be caused by a similar mechanism[244].

Springer and Desai[245] have found that naturally occurring anti-T agglutinins were frequently depressed in patients with adenocarcinoma, as compared with non-carcinomatous persons, and others[246] have reported similar results. Bray et al.[247] confirmed decreased levels of humoral lytic anti-T by cytotoxicity assay in patients with metastatic gastrointestinal cancer, and were able to correlate the extent of depressed anti-T with the extent of tumor progression. It is noteworthy that following surgical resection of primary tumors, many patients developed high anti-T titers[221, 248].

The significant fall in the amount of anti-T antibodies pre-operatively appears to be secondary to the absorption of these antibodies on the surface of tumor cells[197, 221, 248, 249]. The possibility of fluctuating anti-T levels by non tumor related factors has been noted[250].

Dube et al.[251] have reported that B lymphocyte cytotoxins in breast cancer sera were related to red blood cell cold agglutinins (presumably including anti-I) which the authors believed constituted a portion of the cytotoxic immunoglobulins.

Levine[105] reported that a naturally occurring hemolysin reactive against a Forssman-like mucosal tissue globoside was present in 70% of random normal human sera. In contrast, only 30% of cancer sera contained the antibody. Other workers have found that Forssman-like antibody levels were decreased in gastrointestinal, biliary and lung tumor patients in comparison with age matched non-cancer subjects[252, 253].

Deegan et al.[254] studied a man with chronic lymphocytic leukemia whose peripheral blood and splenic lymphocytes contained monoclonal surface Ig, Ia-like antigen and receptors for unsensitized sheep red blood cells. The receptor was not blocked by monoclonal antibodies that bind to the classic T-lymphocytic rosette receptor, but was blocked by anti-human IgM and Kappa antisera and by Forssman antigen. Thus, Forssman-like Ig appeared to be uniquely expressed in this lymphoproliferative disorder. Since this antibody will cross-react with A determinant, its ubiquitous nature in normal individuals and in some cancer patients[105, 209] may provide an immune mechanism for cancer cell elimination.

Naturally occurring antibodies, some with anti-A specificity, were found by Knuth et al.[255] to be directed against blood group determinants on colon cancer cells. The antibodies, present in normals and families of high risk cancer patients, differed in two high risk individuals in that they were directed against A and B antigens on tumor cells but not on red cells.

Exposure to embryonic proteins or precursors of blood group substances was believed by Pompecki et al.[256] to account for unusually high elevations of anti-Le^{ab} antibodies in patients with cancers of breast, lung and colon. In their studies the authors utilized a radioimmunoassay in which labelled immune complexes were precipitated by polyethylene glycol. The authors believed that such antibodies may possess a functional role by blocking T cell killing of tumor cells or by causing lysis via complement action.

Erythrocyte autoagglutinins may occur in malignancy and may occasionally cause active hemolysis[257]. Diseases in which this problem has occurred include chronic lymphocytic leukemia (CLL), and Hodgkin's disease and other lymphomas[258]; in CLL, anti-Rh IgG antibodies and anti-I IgM have been described. Autoagglutinins have also been described in thymoma[259], and Kaposi's sarcoma[260]. Anti-i and anti-A_1 autoagglutinins have been associated with reticulosarcoma[261, 262] and an IgG anti-I^T was described in Hodgkin's disease[263].

Boughton[264] reported on several patients with choriocarcinoma who developed anti-i. One patient with disseminated disease developed very high anti-i in titers which caused an auto-agglutination of her red cells and caused minor blood grouping and crossmatching difficulties but no apparent clinical problems. Anti-N autoantibodies were reported in one case of urinary bladder transitional cell carcinoma[394]. Ovarian tumors in several instances have been associated with red cell autoagglutinins[395–397]. In one case, removal of the tumor resulted in a loss of agglutinins, and lyophilized powder from the tumor was capable of absorbing these atypical agglutinins from serum[397].

Castella et al.[265] described an autoantibody on red cells of a patient with metastatic adenocarcinoma. The patient, previously untransfused was blood group A_1; his serum and red cell eluate contained an IgM immunoglobulin which agglutinated A_1 cells, but not A_2, B or 0 red cells or the patients tumor cells. It was not associated with hemolytic anemia.

The reasons for autoagglutinin formation are not clear, but may be the result of (1) tumor cell modified red cells which become foreign to the host and induce an immune response (2) cross reactions between tumor antigens and naturally occurring antibodies.

B. Cell-Mediated Immunity

The above findings represent manifestations of humoral immunity in cancers. Cell mediated immunity is likewise affected as indicated by the development of delayed type hypersensitivity (DTH) to T-related antigens[177, 213, 267]. Springer et al. have studied an impressive list of cancers, and non-cancer patients and healthy individuals by measuring the skin responses to a preparation of healthy red cell derived T antigen injected into the arm (a negative reading using an extract derived from MN red cells is an important requirement for the interpretation of this test). They observed positive DTH in most but not all adeno small cell and squamous cell carcinomas, and other cancers including melanoma. The organs involved were lung, pancreas, breast, skin, and urinary bladder. Responses to T antigens appear to be made by cancers in all stages, including early disease. Interestingly, operative intervention in tumors which were resectable resulted in a reversion of positive DTH to negative, in contrast to humoral anti-T antibodies which tended to increase in titer postoperatively[221, 248].

6 Blood Group Structure and Biosynthesis in Malignancy

A. Abnormalities of Blood Group Precursors – Novel Substrates

The effects of transformation are:

1. to produce simplified glycolipid or glycopeptide patterns in comparison with untransformed controls[268, 269]; and
2. to synthesize new or unusual forms of glycoconjugate.

There is increasing evidence for the presence of unusual membrane changes and of novel substrates in malignancy[270]. This creates the possibility during tumor development of increased competition between rival receptors, some normal and some tumor-associated. If acceptor specificities of limiting blood group transferases or other intermediates are not strict, terminal sugars may be divided between conventional carriers and those which arise following transformation.

A shifting of glycolipid synthesis from one series to another can be seen upon oncogenic transformation[2]. Thus, decreases in globosides and increases in lactoneotetraosyl ceramides have accompanied the transformation of NIL cells[271]. Such alterations may favor corresponding shifts in P[5, 272] or AB0 and Ii[2, 273] antigenic specificities.

Synthesis of the above macromolecules along with their terminal sugars is profoundly affected by alterations in glycolipid organization which accompany tumor-promoted changes in growth or the cell cycle[2]. For example, ganglioside abnormalities are known to characterize many tumor cell types[274, 279]. Since some gangliosides have been shown to modulate the expression of glycoproteins[275], and a role for ceramide in defining membrane glycolipid has been postulated[276], changes in their amphipathic nature induced by tumor could result in changes in membrane glycoconjugates.

The methods in use for structural analysis include methylation, mass spectrometry, nuclear magnetic resonance[277], stepwise degradation with exoglycosidases[24], examination of oligosaccharides liberated by endoglycosidases[278], and immunostaining of glycolipids separated by thin layer chromotography[201]. Monoclonal antibodies prepared against defined oligosaccharides have been of use in the isolation of a number of naturally occurring novel blood group substances or their precursors[2, 200, 201]. A combination of the above methods has yielded the following structures:

A-like antigen (GalNacβ 1 → 3 Galβ 1 → 4 Hex) without fucose in tumors of group 0 patients[280],

P-like antigen (GalNacβ 1 → 3 Galβ 1 → 4 GlcNacβ 1 → 3 Galβ 1 → 4 Glc) in a tumor of a pp patient[281] and

polyfucosylated type 2 chain (Galβ 1 → 4 Fucα 1 → 3 GlcNacβ 1 → 3 Galβ 1 → 4 Fucα 1 → 3 GlcNacβ 1 → 3 Galβ 1 – R) in a human colon cancer[282],

mono- and disialyl-Lewis structures (gastrointestinal cancer antigens or GICA)[283]

fucosylated 1 → 3 and 1 → 4 GlcNac in neuroblastoma polypeptides[284], and the unique lung tumor related oligosaccharide described by Kabat et al.[211] (see above).

Hybridoma antibodies have recently been prepared by Hakomori et al.[399, 400] against glycolipid purified from human adenocarcinoma tissue. It was possible to select antibodies which differentially recognized mono, di and tri-fucosylated type of chains. The latter two structures were prevalent in colonic and liver adenocarcinoma, but virtually absent in normal colon and liver tissue. Solid-phase radioimmunoassay and a chromatogram binding assay were used by Hanson et al.[389] to characterize the binding specificities of five monoclonal antibodies generated from mice immunized with human tumor cell lines when tested against various glycolipids. Four antibodies derived from mice immunized with pancreatic carcinoma cells detected specifically the human blood group B determinant Galα 1 → 3 Gal (2 ← 1 αFuc). These antibodies preferred type 2 glycolipids (Galβ 1 → 4 GlcNac). An antibody derived following immunization with a human rectal carcinoma cell line possessed binding properties identical to the antibody which reacts with the stage-specific embryonic mouse antigen (SSEA-1) bearing the determinant Galβ 1 → 4 GlcNac (3 ← 1 αFuc)[408].

In studies of cancer-related Forssman structures, Hakomori et al.[104] prepared extracts from gastrointestinal cancers in 21 patients. In 16 Forssman negative patients, some of them blood group 0 or B, the tumor tissue was found to possess Forssman A-like properties, whereas in 5 Forssman positive cases, tumor tissue extracts lacked this property.

Within the P system, P^k antigen which contains an abnormally large quantity of ceramide trihexoside and no detectable globoside, was found to occur more regularly on B lymphocytes from normal and leukemic donors, than on T lymphocytes which tested as P^k negative[286]. Lectin binding studies have shown decreases in receptor sites on leukemic lymphocytes in comparison with normal cells[287].

The occurrence of T asialoganglioside, as well as glycopeptide[288], has been reported in human and guinea pig carcinomas using antisera specific for Galβ 1 → 3 GalNac linkages. Sialylated M positive tissue from normal sources reacted poorly in absorption studies with these sera in contrast with tumor extracts.

B. Abnormalities of Glycosyltransferases

In cases of blocked or incomplete synthesis, the deleted blood group expression has frequently been matched by reductions in the corresponding gene-specified glycosyltransferases[155]. Thus, Stellner et al.[391] demonstrated marked differences in the conversion rate of H_1 to A^a glycolipid when normal versus adenocarcinomatous gastrointestinal epithelia were compared as sources of blood group glycosyltransferase enzymes in a biosynthetic assay. Kim et al.[180] found that in a comparison of normal with carcinomatous colon, losses of A and Le^a antigens, as well as the terminal galactose precursor of H substance, occurred in cancer tissue as, inferred from reduced reactivities of tumor in the presence of *Ricinus communis* lectin. Their data suggested that the absence of A antigen was not only due to lack of terminal GalNac glycoconjugates, but possibly also to failure of tumor tissue to form proper amounts of H determinant, which is a prerequisite to formation of the A structure. The deficiencies in A and H antigens could be related to deficiencies of the corresponding specific transferase enzymes. Similar correlations were reported by us in 23 carcinomas, 14 of which were tissues derived from colon[66].

Although findings by other investigators in colonic and stomach tumors vary from the above in some respects[289)] a frequently cited result in tumor transformation is a reduction of specific sugars on glycolipids, owing to blockades in the synthesis of glycosyltransferases[2, 269)]. The acceptors that are exposed following transformation could themselves possess serologic activity, i.e., T antigen, or they could be novel, but functional as blood group acceptors (see above).

Changes in hepatic fucogangliosides with novel fucosyltransferase have been produced experimentally following tumor induction in the rat by Holmes and Hakomori[290)]. In addition to enhanced α-galactosyltransferase activity in Golgi membranes causing accumulation of a B active α-galactosyl α-fucosyl Gm_1 ganglioside, there was produced a unique α-fucosyltransferase which provided the substrate for the agents described above. The fucosyltransferase possessed physicochemical properties which differed from other tissue transferases, including that responsible for H antigen biosynthesis. The authors suggested that this unique enzyme might serve as a marker in premalignant and malignant states. Several unique tumor related structures mentioned earlier i.e., A-like antigen on a fucoseless substrate, and Forssman A-like antigen await the characterization of specific glycosyltransferases.

In addition to their membrane localization, glycosyltransferase enzymes may also be found in the serum in a number of species, including man[4, 291–293)]. In the case of the human AB0 groups, the glycosyltransferase(s) corresponding to the individual's erythrocytic group are present, i.e., the group A enzyme is N-acetylgalactosaminyl transferase, and the H enzyme is α-2-L-fucosyltransferase. Presumably, these enzymes are released from secretory cells and from disintegrating cells of various types, but little or nothing is known of their exact origin or turnover. The interesting suggestion has been made that they may mediate the repair of altered glycolipids of red cell or other hematopoietic cell surfaces, and thus maintain the function and antigenicity of these cell structures during their lifetime in the circulation[294)]. In any event, the relative ease of measurement of serum enzymes has created much interest in their use as moniters or predictive aids[295)].

In our studies of serum glycosyltransferase enzymes, serial determinants for ABH enzymes by low-molecular-weight acceptor assay were carried out in specimens from patients with AML and ALL[141, 390)]. The samples for study were derived at the time of presentation, during clinical remission, and during the period of relapse. Particularly striking changes during the serial course of the disease were noted in the case of α-L-fucosyltransferase (H enzyme). In practically all presentation sera from untreated patients, abnormally low values (expressed as percentages of radioactive sugar incorporated from nucleotide into product) were demonstrated (Fig. 4), but reverted to normal levels at the time of clinical remission when patients were on combined regimens of chemotherapy; with the development of drug resistance and clinical relapse, enzyme values became low once again (Fig. 5). Negligible changes were observed in the case of A enzymes.

It is not yet known whether these alterations represent decreased enzyme synthesis or increased degradation in untreated and relapsed patients. It is more likely that the alterations represent the former situation. The ubiquity of occurrence makes it appear unlikely that such a change plays a primary role in blood group loss in leukemia, discussed in an earlier section, although such enzyme deficiencies might

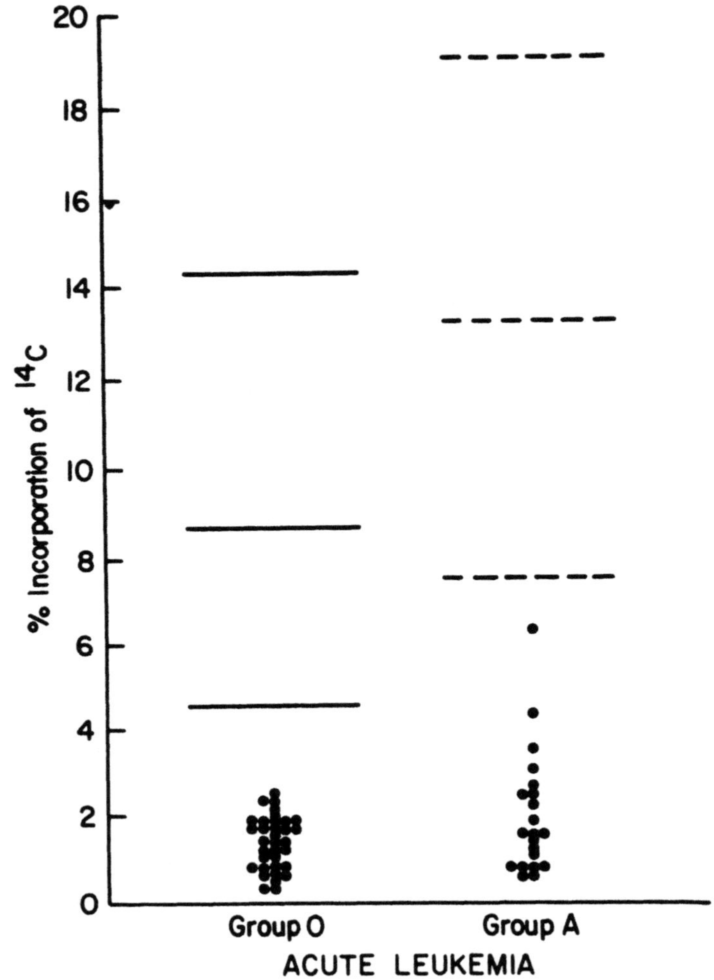

Fig. 4. Serum α-2-L-fucosyltransferase enzyme (H enzyme) in acute leukemia patients of blood groups 0 and A. Serum for H enzyme assays was separated from freshly drawn blood and frozen immediately. Enzyme assays were carried out on freshly thawed samples as follows: 20 µl serum, 5 µl GDP-[^{14}C]fucose (70,000 cpm), 10 µl 5% aqueous phenyl-β-D-galactoside, 5 µl 0.1 M neutral ATP, 5 µl 1% sodium azide, and 30 µl 0.1 M Tris-HCl buffer, pH 7.2, were mixed and incubated for 16 hr at 37 °C. Radioactive products were chromatographed on Whatman 40 paper in a 10 : 4 : 3 mixture of ethyl acetate, pyridine, pyridine and water. With the concentration of acceptor used in the incubation mixture, incorporation of L-[^{14}C]fucose into phenyl-β-D-galactoside is linear up to 16 hr[3]. Horizontal lines, low, mean, and high H enzyme values in normal donors. Group 0 donors, ——; group A donors, – – – (From Kuhns et al.: Cancer Res. *40,* 268 (1980), reprinted with permission)

well contribute to antigen loss. H enzyme is felt to be derived mainly from hematopoietic cells, on the basis of the observation that persons of the para-Bombay phenotype, who lack erythrocytic H antigen but contain secretor substance in their saliva, do not appear to have H enzyme in their sera[298] or at least, low levels of enzyme[297]. Thus, variations in the quantity of serum enzyme may properly be

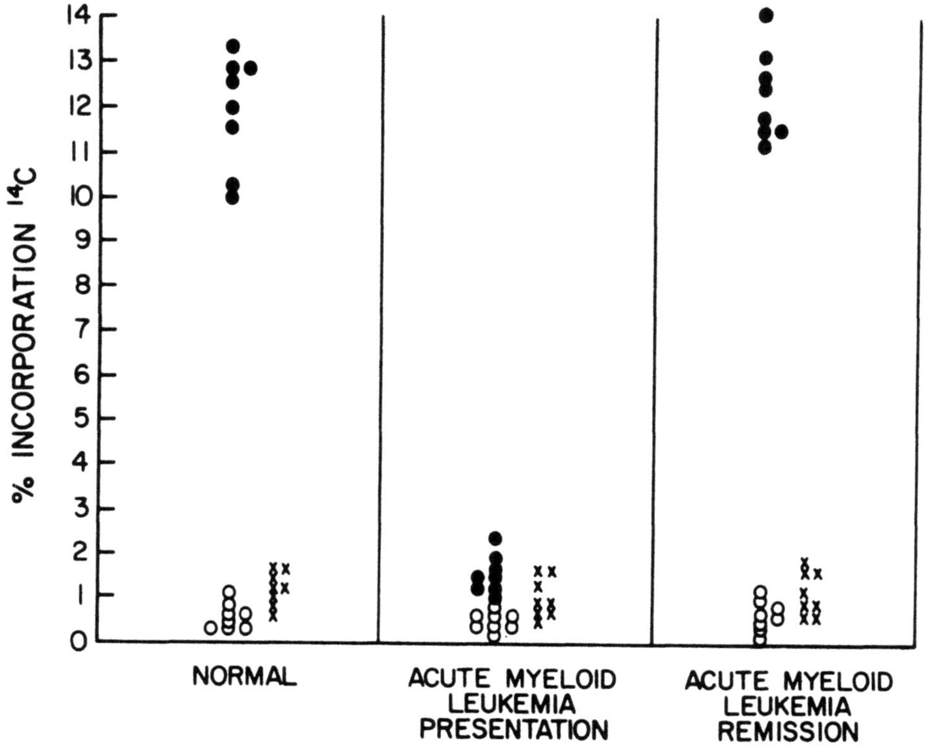

Fig. 5. Serum α-2-L-fucosyltransferase enzyme (H enzyme) compared in normal individuals and in leukemic patients prior to treatment and at the time of remission. Assays were performed as described under Fig. 4. The low molecular weight acceptor assay involves ethanol precipitation of the incubation mixture and detection of sugar-exogenous acceptor product in the supernatant. In this experiment, the low levels of radioactivity in the precipitates indicated that the precipitable portion of the incubation mixture possessed little or no endogenous acceptor. Note that H enzyme values in leukemia patients at presentation were abnormally low but reverted to normal levels at remission. ● ^{14}C product using exogenous acceptor; ○ ^{14}C present in ethanolic precipitates; x ^{14}C product, exogenous acceptor omitted (From Kuhns et al.: Cancer Res. *40,* 268 (1980), reprinted with permission)

related to hematopoietic tissue, and for this reason such variations may be useful as a predictive aid in leukemic therapy. Salmon[156)] has indicated that red cells which demonstrate ABH losses are themselves deficient in the corresponding transferases.

A high-molecular-weight assay for α-2-L-fucosyltransferase was studied in leukemic sera by Khilanani et al.[297)]. The authors measured the levels of a transferase in the plasma of 18 patients with acute adult leukemia at various clinical stages, in conjunction with simultaneous bone marrow aspirations and biopsies. Desialylized fetuin, a high-molecular-weight (HMW) acceptor, was utilized in these enzyme assays, and, on the basis of previous studies involving the use of this acceptor, it was assumed that the enzyme assayed was the H gene-specified α-2-L-fucosyltransferase. Patients in remission, however, had significantly lower levels of this enzyme than did nonresponding or relapsing patients, and elevated plasma enzyme levels in presentation samples could be correlated with the percentage of marrow blast cells. These results are completely at variance with our findings using the low-molecular-weight

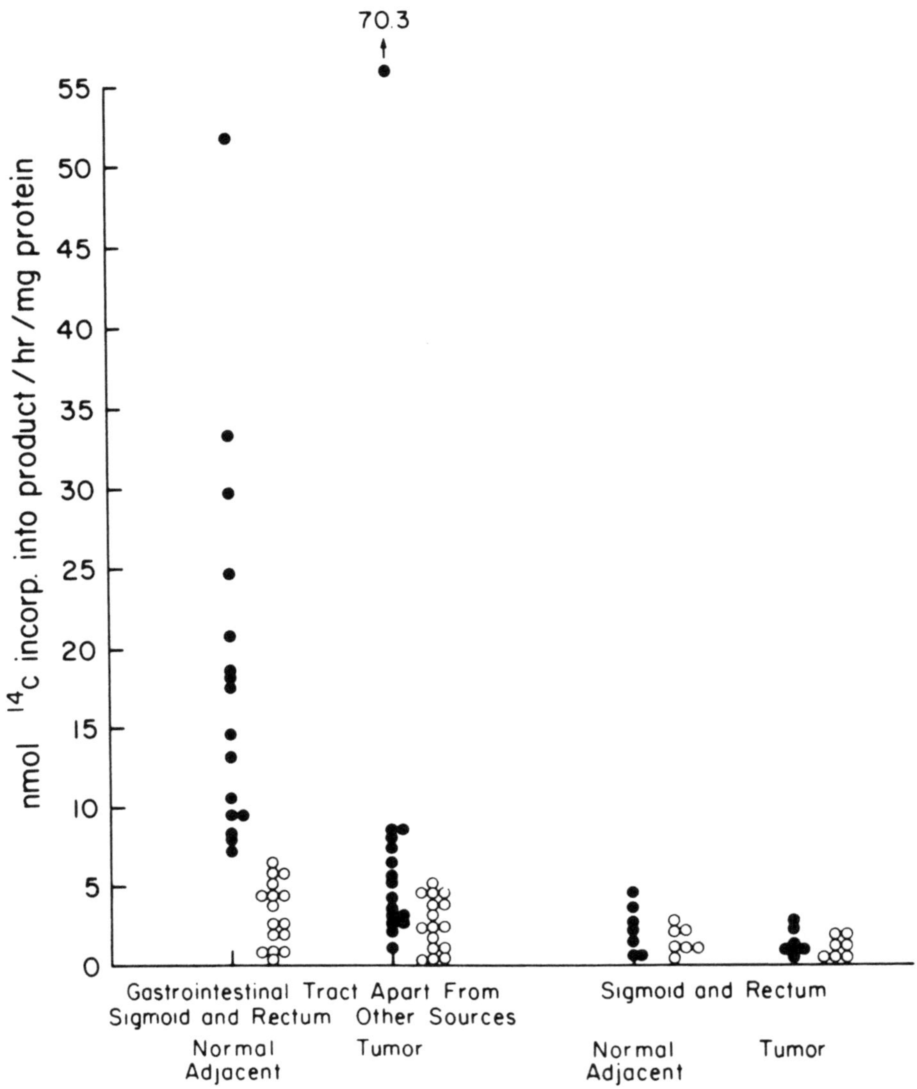

Fig. 6. Comparisons of α-2-fucosyltransferase derived from mucosal extracts of tumor and normal adjacent tissue. Enzyme assays were carried out as described under Fig. 4. H enzyme results in the presence of exogenous acceptor, ●; H enzyme results in the presence of acceptor, ○. Mucosal adjacent and tumor mucosa from sigmoid and rectum both derived from other sources. With one exception (stomach), normal adjacent mucosal from other sources demonstrated high enzyme values relative to the tumor transformed counterpart

acceptor phenyl-β-D-galactoside. It was suggested that competition between endogenous (high-molecular-weight) acceptor and the exogenous (low-molecular-weight) acceptor might account for the discrepancy. Experiments designed to test this hypothesis have, however, failed to reveal competitive endogenous acceptors in sera from leukemic patients when these are assayed with the low-molecular-weight

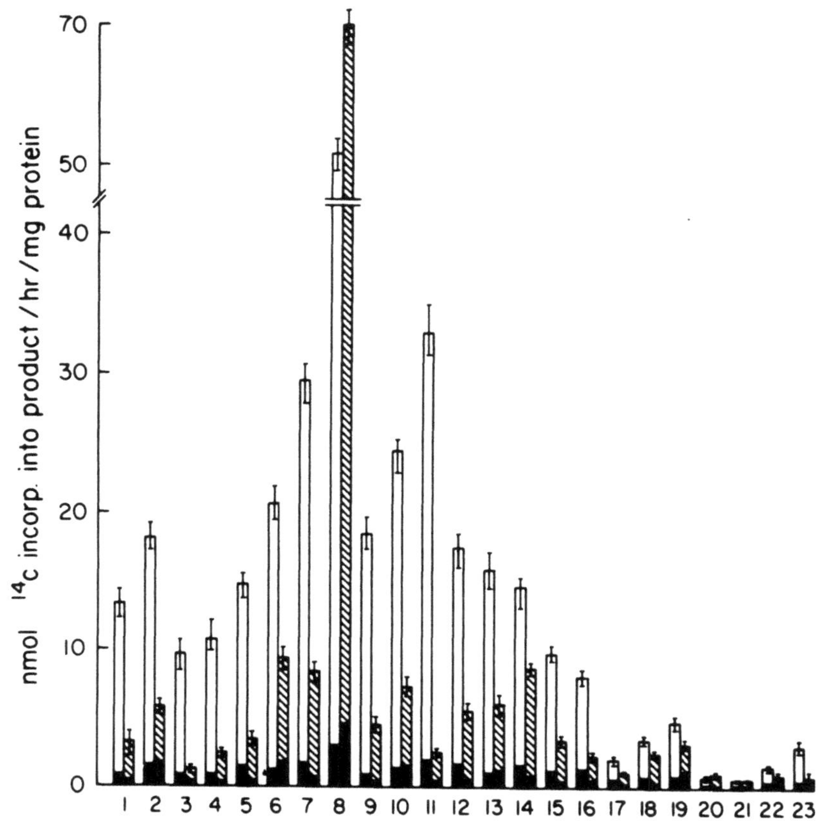

Fig. 7. α-2-L-fucosyltransferase in tumor versus normal adjacent cell extracts. Comparisons of free [^{14}-C]fucose and acceptor-sugar product in chromatographically separated incubation mixtures. ^{14}C-labeled product (fucosylphenyl-galactoside) in normal tissue, □; product in tumor tissue, ▨. Free [^{14}C]fucose is represented in black as a proportion of the total bar area. Each determination carried out on 3 samples. Abscissa, locations of tumor which are enumerated in the table below

Case	Location of carcinoma	Blood group	Case	Location of carcinoma	Blood group
1	Mouth	0	13	Transverse colon	A
2	Mouth	A	14	Transverse colon	0
3	Tongue	A	15	Descending colon	0
4	Larynx	B	16	Descending colon	A
5	Larynx	0	17	Sigmoid	B
6	Esophagus	B	18	Rectosigmoid	A
7	Lung	0	19	Rectosigmoid	A
8	Stomach	0	20	Rectum	0
9	Cervix	0	21	Rectum	0
10	Cecum	0	22	Rectum	A
11	Cecum	0	23	Rectum	A
12	Ascending colon	A			

acceptor. The possibility of increased glycosidase activity, which could result either in cleavage of L-fucose from the product of enzyme transfer, or in a lowering of the acceptor substrate concentration by hydrolysis of phenyl-β-D-galactoside, has been ruled out in our experiments. High-molecular-weight and low-molecular-weight acceptors were later found to be measuring different forms of fucosyltransferase[298)].

When serum glycosyltransferase enzyme assays were carried out in patients with carcinomas, using the low-molecular-weight acceptor assay, the findings were in the low normal range. When comparisons were made between carcinomatous and normal adjacent tissue extracts, however, generally lower than normal tissue blood group enzymes could be demonstrated and were found to accompany deficiencies in tissue AB0 expression (Figs. 6, 7).

C. Excessive Glycosidases

Increased activitiy of glycosidases have been reported for a number of tumors and transformed cell lines[308–310)]. However, to our knowledge, there are no studies that relate losses of blood group determinants to increases in specific glycosidase activity.

D. Masking of Antigens by Excessive Sialic Acid

Sialyltransferase enzymes have been measured in a number of transformed cell lines and appear to be considerably altered in relation to their normal cellular counterparts. This finding bears a direct relationship to levels of sialic acid which are elevated in many transformed cell lines. Increased sialic acid has likewise been observed in certain carcinomas and on leukemic cells[299–303)].

Blood group expression is related to the expression of terminal sialic acid in the case of the MN-related antigens and the En groups[16, 304)]. The observation of Springer indicating that in tumor transformation, sialic acid immunodeterminants for M and N were removed to reveal T antigen, has been demonstrated as well *in vitro* when M or N erythrocytes or purified glycopeptide derivatives were treated with sialidase[305)].

There is evidence that excessive sialic acid may be one mechanism accounting for decreased antigen expression on acute leukemic cells. Kassulke et al.[139)] reported that erythrocytic groups A and H, which were detectable on normal leukocytes, were decreased in strength on leukemic leukocytes [3 of 6 group A, 1 of 5 group 0(H)]. The studies were carried out on mixed leukocyte populations. Sialidase treatment was found to cause re-expression of such diminished activity, thus raising the important point that masking of otherwise competent antigen sites by excessive sialic acid could account for diminished A or H reactivity. Sialidase treated leukemic leukocytes from six of our group 0 patients have confirmed these findings, but have also indicated that leukemic solubilized membranes possess low quantities of α-2-fucosyltransferase in comparison with normal leukocyte extracts[306)]. The observations on neuraminidase-treated cells are supported by the findings of other investigators that tumor transformation is frequently accompanied by increases in sialyltransferase and by increased production of sialic acid. Sialic acid might be capable of steric hindrance of blood group expression by its presence in excess on neighboring molecules.

In the context of Kassulke's findings, it was interesting that MN expression (the result of terminally placed sialic acid) was enhanced on leukemic leukocytes in comparison to normals, but was abrogated following sialidase treatment.

It is possible that masking of blood group receptors may also be accompanied by defective antigen sites; on the basis of our studies of sialidase-treated leukemic leukocytes derived from group 0(H) patients, this appears to be likely. In contrast to normal group 0 leukocytes which were capable of being enzymatically converted to group A or B when incubated with the appropriate transferase and nucleotide sugar, leukemic cells exhibited no or small amounts of activity, which suggested a deficiency or impairment of A and B carrier molecules[306]. Recent studies of normal leukocytes separated into mononuclear cells and granulocytes have shown significant enzymatic differences between these cell types[404]. The authors stress the need for establishing conditions of cell purity when designing comparative studies.

Conversion of precursor blood group oligosaccharides to full AB0 blood group expression may be affected by the content or distribution of cellular sialic acid. Thus, *in vitro* attempts to enzymatically modify untreated Bombay erythrocytes in the presence of group H enzyme (α-2-L-fucosyltransferase) and its specific nucleotide sugar (guanosine diphosphate L-fucose or GDPF) were unsuccessful. However, when such cells were first treated with sialidase, conversion of the cells to group H could be accomplished[307].

E. Relationships of Blood Groups to Carcinoembryonic Antigen and Other Tumor Markers

Carcinoembryonic antigen (CEA) is a glycoprotein of approximately 200,000 molecular size. It is expressed in the embryo, particularly in the large intestines, and postnatally from carcinomas of the colon and their metastases, as well as other cancer types. It can be emitted from primary and metastatic cancer cells, from whence it gains access to the circulation[311–313]. It is also a surface membrane component of these cells[314, 317]. Circulating CEA can be detected in blood sample in nanogram quantities by radioimmunoassay[315]. This technique as well as immunohistochemical studies of large bowel have shown that small amounts of CEA are present in healthy adult subjects[316, 317]. This observation, coupled with findings that proved malignancies occasionally possess CEA in the normal range, somewhat limit the applicability of circulating CEA levels due to reasons of sensitivity and specificity. Its primary current area of application is in the post-operative monitoring of patients to flag possible tumor recurrences or metastases[316, 318]. More recently, labelled anti-CEA antibodies have been utilized for diagnostic imaging and tumor therapy[319].

The question of whether blood groups exist as parts of the CEA structure is controversial. Simmons and Perlmann[320] found that the CEA fraction was identical to the fraction of blood group substances by gel filtration and electrophoresis and by carbohydrate and amino acid composition. Based upon their analytical and periodate oxidation studies, and upon serological tests, they suggested that CEA may be incomplete blood group substances lacking end groups that determine AB0 specificity. This conclusion would be consistent with recent structural studies of Chandrasekaran et al.[321]. Their CEA preparation, as well as those of Simmons and Perlmann were derived from liver metastases. They placed emphasis on the microheterogeneity of oligosaccharide chains which, in tumor derived CEA, exhibited branching predominantly of the tetra-antennary type. The sugar termini consisted of type 2 chains (1 $\rightarrow$ 4 linkages) similar to the preparations of Simmons and Perlmann. Although many chains were sialylated, none possessed terminal blood group sugars despite the accessibility of apparently specific acceptors. Our own immunohistochemical studies of liver metastases as well as some cases described by Wiley et al.[322] also demonstrate deletion of AB0 in metastases derived from antigen-positive tumors. All of these findings appear to agree with those of Gold and Gold[323] whose CEA preparation lacked GalNac the group A immunodeterminant sugar. Paradoxically, the preparations of Simmons and Perlmann did possess GalNac in contrast to those studied by Chandrasekaran et al.

The weight of evidence would support a tenative conclusion that CEA could function as an acceptor for AB0 sugars. However, tumor-derived CEA appears to possess terminal type 2 chains or 1 $\rightarrow$ 4 linkages, and thus could not accept Lewis specific sugars, to produce Le^a or Le^b reactivities. A significant gap in our knowledge is the difficulty of interpreting the significance of the above in relation to normal colonic mucosa whose penultimate linkages are predominantly type 1. Hakomori has noted that normal blood group glycolipids of glandular tissue contain type 1 carbohydrate chains whereas adenocarcinomata contain type 2 chains for the backbone structure in the majority of glycolipids[330]. The functionality of type 2 linkages as ABH acceptors on CEA has been demonstrated in enzymatic modification studies[324].

The positioning of fucose upon the GlcNac of type 2 chains raises the question whether a unique structure such as stage specific antigen[325] may be of importance in tumor CEA expression, although the latter was in fact improved by defucosylation according to the findings of Simmons and Perlmann. On the other hand, experiments by Ormerod[326] and Pompecki[78] support a possible role for fucose in CEA reactivity.

The CEA molecule as well as its blood group relationship has been defined in recent studies utilizing plant lectins[78, 327–329]. Pompecki et al.[78] studied six CEA preparations derived from liver metastases using a panel of lectins and antisera in a polyethylene glycol precipitation assay as a means of defining specific blood group activity. All preparations studied exhibited blood group determinants, although CEA derived from group A patients expressed A well in only one instance similar to findings of Holburn[331] in earlier studies. Lewis activity was a common finding; the common presence of Le^a in CEA derived from liver metastases is inconsistent with our immunohistologic studies of this antigen on metastatic tissue. Perhaps this reflects the relatively undifferentiated status of these cells. We have demonstrated that when this antigen is present in primary tumors, it can also be demonstrated in metastases almost without exception[317]. These results together with the biochemical findings indicate type 1 and 2 chain heterogeneity of carbohydrates within many primary colonic adenocarcinomas and their metastases. It remains to be demonstrated whether and to what extent CEA molecules themselves bear blood group determinants in vivo.

7 Structure Function Relationships Derived from Studies of Cultured Cells

Blood group erythrocytic antigens have been used as classic markers in studying human population genetics. Somatic cell antigens have been studied by techniques of mixed agglutination[332)], immunofluorescence[333)]; ferritin-EM[334)] and immunoperoxidase[22)]. The finding of AB0(H) and other blood group antigens on the surfaces of cultured somatic cells has made it possible to apply tissue culture methodology to these observations and to utilize these markers for genetic and metabolic studies[335)].

Kelus et al.[336)] first demonstrated the presence of 0(H) antigen on HeLa cells by mixed agglutination, estimating rosettes formed in the presence of anti-H ulex reagent and group 0 erythrocytes. Högman[337)] used the same technique and the appropriate specific reagents to define ABH blood group antigens on primary cell cultures. An established line (FL) was found by Hagiwara[338)] to possess blood group B. Chessin et al.[339)] confirmed the findings of Kelus et al. and Coombs. They also attempted to maintain or enhance *in vitro* blood group activity by addition of constituent blood group sugars or amino sugars to culture media but were not consistently successful in this regard. The earlier literature on blood group detection on cultured somatic cells has been reviewed[340)].

It was noted that 0(H) antigenicity was maintained best by cells in culture, and that A, Tj^a, and I-i antigens, if present at the time of initial culture, tended to diminish in strength following successive cell passages[341)]. Contamination with other cell types can complicate this problem, and mixed agglutination has in fact been useful in identifying components of mixed cell populations, on the basis of unique surface markers[342)]. However, in carefully studied primary cells it did appear that in relation to 0(H) antigens, A and B surface blood group markers were relatively labile, and may have been lost from cell membranes as a result of the altered cell growth customarily encountered during attempts to maintain primary cultures[343)]. The transformed cell cultures tested by us have demonstrated 0(H) activity, but none of the other antigens of this system[339)]. Changes in cell organization may cause blood group alterations even when a blood group can be maintained in continuous culture. Forssman antigen was alternatively induced and repressed, depending upon whether separated cell were grown in vitro, or reaggregated and grown by an organ culture technique[344)].

Fellous et al.[208)] have utilized a complement-fixation technique to demonstrate blood group antigens on fibroblasts and, confirming other authors, noted long-term instability following multiple cell passages. The specificities detected by these workers on cells as far as the 13th passage included H, I, penumococcus XIV, and Le^b; however, A antigen was lost as early as the second or third passage.

On the other hand, Jones et al. reported that A antigen, as well as CEA and a colon specific mucin were retained in colon carcinoma line HT-29 assayed over more than 170 generations[345]. It would be of interest to determine whether the A precursor possessed H antigen or some other unique fucoseless structure. Trypsin and other proteolytic enzymes have been shown to enhance 0(H) antigen expression, presumably by removing surface macromolecules which hinder glycolipid expression of this immunodeterminant[346]; this method was used by Katoh et al.[347] to enhance ABH reactions on brouchial epithelial cells.

Momota et al.[348] used Epstein-Barr virus to transform and establish lymphoid cell lines from persons with different P blood group phenotypes. Thin layer chromatography was used to examine the glycolipid composition of these lines. P_1^k and P_2^k phenotypes were found to lack globoside (P antigen), the p phenotype was deficient in both globoside and ceramide trihexoside (P^k), whereas both glycosphingolipids were present in cell lines which possessed the usual P_1 or P_2 phenotypes. CTH synthetase was detected at decreased levels in p cell lines. Genetic complementation studies have been carried out using hybrids prepared from p and P^k fibroblasts[208].

Some epithelial cells in culture appear to spontaneously shed M, N, T, and Tn antigens, which can be detected in culture supernatant fluid. Thus B0T-2, a human mammary tumor-cell line, was capable of forming serologically reactive glycoproteins[349]. An N-like glycoprotein has also been found in the serum and ascites fluid of mice bearing the H_a subline of TA_3 murine mammary adenocarcinoma[350, 351]. A blood group H glycolipid obtained from rat ascites hepatoma cell line HH 7974F has been reported[352]. Other non-human sources of H glycolipids include the dog[353, 354], pig[335], and cow[356].

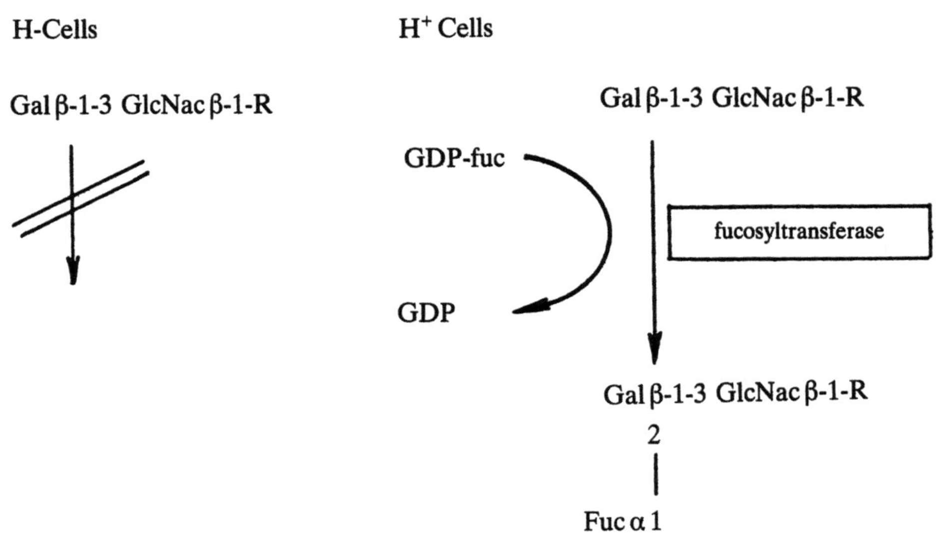

Fig. 8. A suggested molecular model for the expression of blood group H on Hela cells. When group H is expressed, a fucosyltransferase enzyme is activated and thus enables a sugar nucleotide (GDP-fucose) to attach the corresponding sugar to the preexisting macromolecule (From Kuhns et al.: Am. J. Path. *69,* 389 (1972), reprinted with permission)

The ability to define H antigen on HeLa cells cultured under various conditions suggested the use of these cells for studying basic processes concerned with phenotypic expression of a blood group[357]. HeLa cells were derived originally from an adenocarcinoma of the cervix, and since their transformation in 1952 have been successfully passaged in cultures for numerous generations[358]. A number of strains have been established as the result of cloning, and enzyme or marker characteristics vary somewhat depending upon the strain used[358]. Thus far, all HeLa cell strains tested by us have yielded positive results for H antigen by mixed agglutination, immunofluorescence, or ferritin EM.

The findings of mixed H+/H-cell populations prompted us to carry out cloning and population studies which demonstrated that H-cells could function as stem cells for the production of mixed H+/H-progeny[359]. Chemical synchrony using thymidine showed that interphase populations possessed high proportions of H-cells, whereas mitotic cells were primarily H+[359, 360]. Tests for H-gene specified α-2-L-fucosyltransferase showed primary production or activation during the phase of synthesis and early G2[361]. Fucosyltransferase enzymes have been isolated from HeLa cells by other investigators[362]. The presence of H gene product, in addition to its phenotype expression (Fig. 7) as well as serologic demonstrations of P system components provide a unique opportunity to study translation patterns of membrane receptors in relation to their growth and cell cycle relationships. Isotope labelling studies using fucose or glucose incorporation in cultured cells may be helpful in this regard[363–365].

Fucolipid and fucoprotein alterations in tumor transformation have been demonstrated[366–369]. Possible relationships between tumorigenecity and glycosphinglolipid glycosyltransferases in cultured transformed cells have been explored and the literature discussed by Basu and Basu[370]. Among the alterations cited are novel acceptors for blood group sugars accompanied by abnormal levels of transferases[371], as well as unusual expression of such enzymes in Forssman glycolipid biosynthesis[372].

8 Discussion and Conclusions

Present knowledge of the biochemical genetics of the major blood group system permit reasonable interpretations of how derangements in these metabolic pathways could be related to cancer if somatic mutation is presumed as a primary event. Bacterial somatic mutation has provided an ideal model for studying the role of the bacterial nucleus in controlling the synthesis of cell surface antigens through a coordinated sequence of regulating enzyme systems. Moreover, these studies have also shown how the genetic changes that occur in spontaneous mutations or lysogenic conversion modify normal biosynthetic pathways to produce precisely characterized changes in the chemical structure and hence in the serologic specificity of bacterial surface 0 antigen[373]. Analogous biochemical and serologic conversion appear to occur as a concomitant of mammalian cell tumor transformation, but most likely the translation patterns are far more complicated than tose found in bacterial systems.

Blood group antigens, especially those of the AB0(H), Lewis, Ii, MN, T, and P groups consist of sugars with representation on mucins and on membrane lipids.

Since marked changes in the latter can be induced by cancer, it is not surprising to observe frequent alterations in blood group determinants. The fidelity of expression of the gene-determined glycosyltransferases, and therefore of the individual sugars which they specify on oligosaccharide chains, is normally variable[374)], and can account for the microheterogeneity which is characteristic of glycolipids and glycoproteins[409)]. However, loss or change in regulation of only one gene coding for a glycosyltransferase enzyme can produce marked alterations in an oligosaccharide sequence, depending upon the position of such sugars in the sequence of expression, whether terminal or proximal. Loss of a blood group in cancer could thus be accounted for if its specific transferase were no longer being coded for, if a similar block occurred early in the sequence of sugars bearing a terminal blood group antigen[375)], or if major shifts occurred in a cell population whose membranes did not possess such a marker or its enzyme. It may also be of importance that glycolipid alterations in cancer cell membrane may possess a modulating influence upon protein domains or the glycosylation of proteins[376)].

Apart from marker functions, there is no way of telling how uniquely important blood group changes are in cancer development, or what they signify. The question has been raised whether certain Lewis phenotypes can place individuals at risk for gastrointestinal cancers[199)].

Stomach and colon carcinomas are unique in their apparent tendency to produce two patterns:

1. antigen loss; and
2. gain of A-like Forssman antigen.

The latter has been termed nongenetic by some investigators[188)], but Levine[105)] views both as components of a phenotypic system. The changing blood group composition of normal colon from embryo to adult should also call attention to the possibility that reexpression of embryonic blood group antigens may occur in carcinoma in the adult[45, 187)]. Changes in bacterial flora, which are known to occur in gastrointestinal cancer, for example, could play an etiologic role in blood group antigen change[377, 378)]. Some gastrointestinal bacteria possess Forssman-like properties[379)], as well as other blood group like antigens[380)] and presumably have the corresponding transferase enzymes responsible for catalyzing specific sugar additions to mammolian membrane. Whether such enzymes could catalyze sugar additions to human tissue acceptors would need to be determined. Similarily, the possible role of specific microbial glycosidases in blood group loss requires further study.

In the case of hematopietic cells, cultured leukemic cell lines have been helpful in studies of relationships of blood group expression to cellular maturation[45, 144, 173)]. Loss of a blood group phenotype on erythrocytes in leukemia appears to occur less regularly than does blood group alteration in carcinoma tissue. However, loss or reduction of H enzyme seems to occur in most presentation cases of acute leukemia[141, 381)]. It is possible that this could be related to the finding referred to earlier, that proportions of leukemic cells are arrested in the G_0 or G_1 stage of the cell cycle[382)], and in such cases a direct relationship of cell cycle to blood group nonexpression could be postulated. Presumably, the degree of gene expression would depend upon mechanisms which place such cells in cycle, and on the extent of the cell mass involved. The type of cell mass involved might also be of importance in this regard. The lowered serum transferase in acute leukemia may reflect a specific cellular deficiency in the leukemic cell population, but not necessarily in erythrocyte precursors or their differentiated progeny. Altered phenotypic expression of AB0 antigens might be variably influenced by abberrant transferase enzyme synthesis, as well as by physical events which occur in cytokinesis. Chromosomal alterations involving chromosome 9 which carries the AB0 genes have been reported in leukemias[383)].

Membrane lipid and cholesterol alterations have been described in leukemia[384)], and require evaluation as possible influences upon changed antigen expression. Masking of antigens has been suggested as being due to abnormal sialomucins[139, 306)], but the importance of this factor in relation to other cancer-induced cell changes requires further accurate definition.

The likelihood that blood group antigens that are normally masked or unexpressed become revealed raises the question of whether antigens or vaccines prepared from tissues containing such determinants will provide a future mode of immunotherapy. Bird has suggested that blood-group-specific lectins, such as anti-T derived from the peanut, may be useful as specific cytotoxic agents[257)]. The potential of radiolabelled peanut agglutinin as a tumor localizing agent has been investigated by Zabel et al.[385)] using a mouse lymphoma known to bind fluorescein labelled PNA *in vitro*. The radioiodinated lectin showed good tumor localization and rapid blood clearance. Clear images of tumor were obtained in serial scintigraphic imaging by 24 and 48 h. No blood background subtraction was necessary.

The stage specific embryonic antigen-1 (SSEA-1) which expresses an Le isomer is an effective target for tumor localization when using radiolabeled monoclonal anti-SSEA-1 IgM antibody. Specificity for tumors appears to be unnecessary for localization and perhaps drug targeting with this antibody which also reacts *in vitro* with normal tissues[398)]. Additional studies will be required, i.e., such as differences in antibody metabolism, to determine causes for increased retention in tumors rather than in similarly antigenic normal tissues[398)].

Immunologic surveillance is altered following tumor transformation; alternatively, a predisposition to the latter may be induced by primary changes in immune surveillance[386)]. This frequently is reflected by immunoglobulin changes, as indicated earlier, and also by specific alterations in certain naturally occurring blood group antibodies. At times, autoimmune responses which affect red cells appear to be characterized by antibodies with blood group specificities. Some of these can be considered as potential diagnostic or predictive aids if firm relationships can be established with clinical and morphologic alterations in cancer.

Additional precise studies are required for unique correlations of blood groups with cancer;

1. the sequence of normal cellular morphologic maturation must be closely correlated with blood group development as a basis for evaluating abnormal findings;
2. the importance of masking by sialic acid must be correctly assessed in relation to blood group nonexpression both in tumor transformation and in normal cell differentiation;
3. precise forms of chemical definition and quantitation must be developed so that a serologic determinant can be expressed in terms of its biochemical equivalent;
4. *in vitro* cell separation technology must be applied so that the blood group

patterns of tumor cells and their variants, and those of their normal undifferentiated and mature counterparts, can be separately distinguished;
5. the possible role of infectious agents in cancer cell surface alteration requires study
6. further serial patient studies are needed to determine the diagnostic or prognostic usefulness of cancer-induced blood group or blood group enzyme changes, or the usefulness of alterations in naturally occurring antibodies; and
7. blood group-like tumor antigens require further evaluation as potential aids in diagnosis and treatment.

9 Abbreviations

Gal	Galactose	GDP-Fuc	Guanosine diphosphate fucose
GalNAc	N-acetyl galactosamine	GICA	Gastrointestinal cancer antigen
GlcNAc	N-acetylglucosamine	CEA	Carcinoembryonic antigen
Fuc	Fucose	ALL	Acute lymphocytic leukemia
NeuAC	N-acetylneuraminic acid (Sialic acid)	CLL	Chronic lymphocytic leukemia
		AML	Acute myelocytic leukemia
GSL	Glycophospholipid	CML	Chronic myelocytic leukemia
Gal-GlcNAc	N-acetyllactosamine	CDH	Ceramide dihexoside
UDP-Gal	Uridine diphosphate galactose	CTH	Ceramide trihexoside
UDP-GlcNAc	Uridine diphosphate N-acetylglucosamine		

10 References

1. Baldwin, R.: Adv. Cancer Res. *18,* 1 (1973)
2. Hakomori, S. and Kannagi, R.: J. Natl. Cancer Inst. *71,* 231 (1983)
3. Warren, L.: Role of Carbohydrates Bound to Proteins in Biomembranes *11,* pp. 53–78 (A. Nowotny Ed.), Plenum, New York, 1983
4. Watkins, W.: Biochemistry of the AB0, Lewis and P Blood Group Systems cAdv., Human Genet. *10,* pp. 1–136 (H. Harris and K. Hirschhorn (Eds.)), Plenum, New York, 1980
5. Marcus, D.: Sem. Hemat. *18,* 63 (1981)
6. Hakomori, S.: Sem. Hemat. *18,* 39 (1981)
7. Anstee, D.: Sem. Hemat. *18,* 13 (1981)
8. Mourant, A., Kopec, A. and Domaniewska-Sobczak, K.: Blood Groups and Diseases, Oxford Univ. Press, New York, 1978
9. Bird, G. W. G.: Blood Groups: Determinants of Recognition and Susceptibility to Disease, in: Blood Groups and Diseases (G. Garratty (Ed.)), Amer. Ass. of Blood Banks Monograph, 1983
10. Marchesi, V., Ginsburg, V., Robbins, P. and Fox, C. F. (Eds.): Cell Surfaces and Biological Recognition, Alan Liss, New York 1977
11. Kuhns, W.: Blood Group Alterations, in Cancer, Vol. 1, pp. 151–200, in Contemporary Hemat/Onc. (Gordon, A., Lo Bue, J., Silber, R. and Muggia, F. (Eds.)), Plenum, New York, 1980

12. Coon, J. and Weinstein, R.: Blood Group Antigens in Tumor Cell Membranes, Chapt. 7, pp. 173–205, in: Biomembranes, Vol. 11 (A. Nowotny, (Ed.)), Plenum, New York, 1983
13. Salmon, C. H.: Blood Group Antigens and Malignancy, pp. 37–44, in: 15th Congr. Internat. Soc. Blood Transfusion, Paris, 1978
14. Garratty, G. (Ed.): Blood Groups and Disease, Amer. Ass. Blood Banks Monograph 1983
15. Steane, E. (Ed.): Blood Groups and Disease, Amer. Ass. Blood Banks Monograph, 1977
16. Race, R. and Sanger, R.: Blood Groups in Man, 6th Ed., Blackwell, Oxford, 1975
17. Bach, J.-F.: Vox Sang. *45,* 166 (1983)
18. Lemieux, R.: Chem. Soc. Rev. *7,* 423 (1978)
19. Morgan, W. T. J. and van Heynigen, R.: Brit. J. Exp. Path. *25,* 5 (1944)
20. Matsukura, Y.: Vox Sang. *31,* 321 (1976)
21. Coombs, R., Bedford, D. and Rouillard, L.: Lancet *1,* 461 (1956)
22. DeLellis, R., Sternberger, L., Mann, R., Bank, P. and Nakane, P.: Am. J. Clin. Path. *71,* 483 (1979)
23. LePendu, J., Oriol, R., Lambert, F., Dalix, A. and Lemieux, R.: Vox Sang. *45,* 421 (1983)
24. Morgan, W. T. J.: Proc. Roy. Soc. Biol. *151,* 308 (1960)
25. Ginsburg, V.: Adv. Enzymol. *26,* 35 (1964)
26. Schachter, H., Michaels, M., Crookston, M., Tilley, C. and Crookston, J.: Biochem. Biophys. Res. Comm. *45,* 1011 (1971)
27. Schachter, H., Michaels, M., Tilley, C., Crookston, M. and Crookston, J.: Proc. Natl. Acad. Sci. *70,* 220 (1973)
28. Marcus, D. and Cass, L.: Science *164,* 553 (1969)
29. Marsh, W. L.: Brit. J. Hematol. *7,* 200 (1961)
30. Henri, A., Testa, U., Tonthat, H., Riou, J., Titeux, M., Vinchenker, W., Fenihade, F., Galacteros, F.: Am. J. Hematol. *9,* 161 (1980)
31. Hakomori, S., Watanabe, K. and Laine, R.: Pure Appl. Chem. *49,* 1215 (1977)
32. Watanabe, K., Hakomori, S., Childs, R. and Feizi, T.: J. Biol. Chem. *254,* 3221 (1979)
33. Neimann, H., Watanabe, K., Hakomori, S. et al.: Biochem. Biophys. Res. Commun. *81,* 1286 (1978)
34. Feizi, T., Childs, R., Watanabe, K. et al.: J. Exp. Med. *149,* 975 (1979)
35. Hanfland, P., Egge, H., Dubrowski, J. et al.: Isolation and Partial Characterization of a Complex Homogeneous Glycosphinogolipid from Rabbit Erythrocytes having I- and B-like Activity, p. 520: in Schauer, R. (Ed.): Proc. 5th Int. Symp. Kiel, Stuttgart, G. Thieme 1979
36. Springer, G. and Yung, H.: Immunochem. *14,* 497 (1977)
37. Friedenreich, V.: The Thomsen Hemagglutination Phenomenon, p. 128, Levin and Munksgaard, Copenhagen, 1930
38. Uhlenbruck, G.: Zbl. Bakt. *179,* 155 (1961)
39. Springer, G.: Bact. Rev. *27,* 191 (1963)
40. Moreau, R., Dausset, J., Bernard, J. and Moullec, J.: Bull. Soc. Med. Hop., Paris, Seance May 17, 569, 1957
41. Bird, G. W. G.: Vox Sang. *9,* 748 (1964)
42. Bird, G. W. G. and Wingham, J.: Brit. J. Hemat. *23,* 759 (1972)
43. Bird, G. W. G., Wingham, J., Seger, R. and Kenny, A.: Bl. Transf. Immunochem. *25,* 215 (1982)
44. Springer, G., Cantrell, J., Desai, P. and Tegtmeyer, H.: Clin. Immun. Immunopath. *22,* 9 (1982)
45. Szulman, A. J.: Exp. Med. *111,* 785 (1960)
46. Kent, S. P.: J. Histochem. Cytochem. *12,* 59 (1964)
47. Hartmann, G.: Group Antigens in Human Organs, Munksgaard, Copenhagen, 1941
48. Wiener, A.: Blood Groups and Transfusion, 3rd Ed., Thomas, Springfield, 1943
49. Hakomori, S.: Progr. Biochem. Pharmacol. *10,* 167 (1975)
50. Rouger, P., Goossens, D., Gane, P and Salmon, C.: pg. 101 in Blood Groups and Other Red Cell Surface Markers in Health and Disease, Masson, New York, 1983
51. Szulman, A.: Develop. Biol. *14,* 127 (1980)
52. Glynn, L. and Holborow, E.: Brit. Med. Bull. *15,* 150 (1959)
53. Boorman, K. and Dodd, B.: J. Path. Bact. *55,* 329 (1943)
54. Childs, R., Kapadia, A. and Feizi, T.: Eur. J. Immunol. *10,* 379 (1980)

55. Martensson, E.: Glycosphingolipids of Animal Tissue, in (Holman, R. (Ed.)) Prog. Chem. Lipids, Permagon, New York, 1969
56. Lalezari, P., in: The Granulocyte, Function and Clinical Utilization, p. 209 (Greenwalt, T. and Jamieson, G. (Eds.)), Alan Liss, New York, 1977
57. Lalezari, P.: Transplant Proc., 12 (suppl. 1), 12 1980
58. Lewis, J., Draude, J. and Kuhns, W.: Vox Sang. *5,* 434 (1960)
59. Kelton, J., Hamid, C., Aker, S. and Blajchman, M.: Blood *59,* 980 (1982)
60. Brody, J., Beizer, L. and Mobarak, M.: J. Lab. Clin. Med. *65,* 571 (1965)
61. Rosenfelder, G., Herbst, H. and Braun, D.: FEBS. Lett. *114,* 213 (1980)
62. Childs, R. and Feizi, T.: Biochem. Biophys. Res. Comm. *102,* 1158 (1981)
63. Gahnberg, C. and Anderson, L.: Membrane Glycoconjugates in the Maturation and Activation of T and B Lymphocytes, p. 231: in The Glycoconjugates (Horowitz, M. (Ed.)), Academic Press, New York, 1982
63a. Shumak, K., Rachkewich, R., Crookston, J.: Nature NB *231,* 148 (1971)
64. Hiramoto, R., Smith, E., Ghanta, V., Shaw, J and McKibbin, J.: J. Immunol. *110,* 1037 (1973)
65. Wiley, E., Murphy, P., Mendelsohn, G. and Eggleston, J.: Am. J. Clin. Path. *76,* 806 (1981)
66. Kuhns, W. and Schoentag, R.: Cancer Res. *41,* 2767 (1981)
67. Hounsell, E. and Feizi, T.: Med. Biol. *60,* 227 (1982)
68. Primus, F. J. and Kuhns, W.: Unpublished observations
69. Friedhoff, F. and Kuhns, W.: Transfusion *8,* 244 (1968)
70. Dahr, W., Uhlenbruck, G. and Bird, G. W. G.: Vox Sang. *27,* 29 (1974)
71. Krusius, T. and Finne, J.: Eur. J. Biochem. *84,* 395 (1978)
72. Finne, J. and Krusius, T.: FEBS Lett. *66,* 94 (1976)
73. Loomes, L., Uemura, K., Childs, R., Paulson, J., Rogers, G., Scudder, P., Michalski, J., Hounsell, E., Taylor-Robinson, D. and Feizi, T.: Nature *307,* 560 (1984)
74. Fredman, P., Richert, N., Magnani, J., Willingham, M., Pastan, I. and Ginsburg, V.: J. Biol. Chem. *258,* 11,206 (1983)
74a. Pereira, M. and Kabat, E.: J. Cell Biol. *82,* 185 (1979)
74b. Morstyn, G., Nicola, N. and Metcalf, D.: Blood *56,* 798 (1980)
75. Mollison, P.: Blood Transfusion in Clinical Medicine, 6th Ed., Davis, Philadelphia, 1979
76. Hogman, C.: Vox Sang. *4,* 12 (1959)
77. Kelus, A., Gurner, B. and Coombs, R.: Immunol. *2,* 262 (1959)
78. Pompecki, R., Shively, J. and Todd, C.: Cancer Res. *41,* 1905 (1981)
79. Weinstein, R., Coon, J., Alroy, J. and Davidsohn, I.: Tissue Associated Blood Group Antigens in Human Tumors, p. 239: in Diagnostic Immunohistochemistry (Delellis, R. (Ed.)), Masson, New York, 1981
80. Kohler, G. and Milstein, G.: Nature *256,* 495 (1975)
81. Ziegler, A.: Blut *41,* 1 (1980)
82. Renart, J., Reiser, J. and Stark, G.: Proc. Nat. Acad. Sci. USA *76,* 3116 (1979)
83. Nudelman, E., Hakomori, S., Kannagi, R. et al.: J. Biol. Chem. *257,* 12,752 (1982)
84. Kannagi, R., Nudelman, E., Levery, S. and Hakomori, S.: J. Biol. Chem. *257,* 14,865 (1982)
85. Magnani, J., Brockhaus, M., Smith, D. et al.: Science *212,* 55 (1981)
86. Nairn, R. C.: Fluorescent Protein Tracing, 4th Ed., Churchill Livingstone, Edinburgh, 1976
87. DeLellis, R. (Ed.): Diagnostic Immunohistochemistry, Masson, New York, 1981
88. George, D., Hanks, C., Courtney, R. and Lopatin, D.: J. Dent. Res. *59,* 831 (1980)
89. Dabelsteen, E., Vedtofte, P., Hakomori, S. and Young, W.: Cancer Res. *43,* 1451 (1983)
90. Mourant, A. E.: Blood Groups and Disease, p. 159, in: Functions of the Blood (Robb-Smith, A. H. T. and MacFarlane, R. G. (Eds.)), Academic Press, New York, 1961
91. Mourant, A.: Distribution of Human Blood Groups, Thomas, Springfield, 1954
92. Clarke, C. A.: Prog. Med. Genet. *1,* 81 (1961)
93. Roberts, J. A.: Brit. Med. Bull. *15,* 129 (1959)
94. Aird, I., Bertall, H. H. and Roberts, J. A. F.: Brit. Med. J. *1,* 799 (1953)
95. Clarke, C. A., Edwards, J., Haddock, D., Howell-Evans, A., McConnell, R. and Sheppard, P.: Brit. Med. J. *2,* 725 (1956)
96. Vogel, F.: Am. J. Hum. Genet. *22,* 464 (1970)
97. Langman, M. J. S. and Doll, R.: Blut *6,* 270 (1965)

98. Mourant, A., Kopec, A. and Domainiewska-Sobczak, K.: Lancet *1*, 223 (1971)
99. Dausset, J. and Svejgaard, A.: HLA and Disease, Munksgaard, Copenhagen, 1977
100. Aird, I., Beutall, H., Mekigan, J. and Roberts, J. A. F.: Brit. Med. J. *ii*, 315 (1954)
101. Lehmann-Facius, H.: Ergeb. Hygiene *56*, 464 (1928)
102. Hirszfeld, L., Halber, W. and Laskowski, J.: Z. Immunitätsforsch., Immunobiol., I and II, *64*, 61 (1929)
103. Witebsky, E.: Klin. Wochenschr. *9*, 58 (1930)
104. Hakomori, S., Wang, S. M. and Young, W.: Proc. Natl. Acad. Sci. USA *74*, 3023 (1977)
105. Levine, P.: Sem. Oncol. *5*, 25 (1978)
106. Kijimoto-Ochiai, S., Takahaski, W. and Makita, A.: Japan J. Exp. Med. *51*, 149 (1981)
107. Gunz, F. and Henderson, E.: Leukemia, Grune and Stratton, New York, 1983
108. Bagshawe, K. and Rawlins, G.: Lancet *1*, 553 (1971)
109. Bagshawe, K.: Adv. Cancer Res. *18*, 231 (1973)
110. Newton, R., Buchler, S., Crumely, J. and Marshall, W.: Vox Sang. *37*, 158 (1979)
111. Wiener, A. S. and Gordon, E. B.: Brit. J. Haematol. *2*, 305 (1956)
112. Van Loghem, J. J., Dorfmeier, H. and van der Hart, M.: Vox Sang. *2*, 16 (1957)
113. Stratton, F., Renton, P. H. and Hancock, J. A.: Nature *181*, 62 (1958)
114. Salmon, C., Dreyfus, B. and Andre, R.: Rev. Hematol. *13*, 148 (1958)
115. Salmon, C., Andre, R. and Dreyfus, B.: Rev. Fr. d'Etude Clin. Biol. *4*, 468 (1959)
116. Gold, E. R., Tovey, G. H., Benney, W. E. and Lewis, F. J. W.: Nature (London) *183*, 892 (1959)
117. Tovey, G. H.: Proc. 7th Conf. European Soc. of Haematology, London, Karger, 1960
118. Bhatia, H. M. and Sanghvi, L. D.: Indian J. Med. Res. *14*, 534 (1960)
119. Salmon, C.: Ser. Hemat. *II*, 3 (1969)
120. Ayres, M., Salzano, F. M. and Ludwig, O. K.: J. Med. Genet. *3*, 180 (1966)
121. Salmon, C. and Salmon, D.: Rev. Fr. Etudes Clin. Biol. *10*, 212 (1965)
122. Kurokawa, H., Katsura, S., Kamadu, M., Matsuo, W., Tokhwa, Satodate, R. and Sasaki, K.: Tohoku J. Exp. Med. *99*, 35 (1969)
123. Ley, A. B., Harris, J. P. and Brinkley, M.: Proc. Amer. Assoc. Blood Banks, p. 72, 1961
124. Undevia, J. V., Bhatia, H. M., Sharma, R. S. and Parekh, J. G.: Indian J. Med. Res. *54*, 1145 (1966)
125. Salmon, C., Salmon, D., Micorin, C. and Berthier, J.: Rev. Fr. Hematol. *6*, 563 (1966)
126. McGinniss, M., Schmidt, P. and Carbone, P.: Nature *202*, 606 (1964)
127. Schmidt, P., Barille, M. and McGinnis, M.: Nature *205*, 371 (1965)
128. Shumak, K., Rachkewich, R., Beldotti, L.: Brit. J. Hematol. *41*, 407 (1979)
129. Reviron, J., Schenmitzler, C., Weisgerber, C. and Bernard, J.: Rev. Fr. Transfus. *13*, 251 (1970)
130. Majsky, A. and Brabec, V.: Folia Haematol. *78*, 237 (1961)
131. Ayres, M., Salzano, F. M. and Ludwig, O. K.: Acta Haematol. (Basel) *31*, 150 (1967)
132. Tovey, G. H., Lockyer, J. W. and Tierney, R. B. H.: Vox Sang. *6*, 628 (1961)
133. Ruffie, J., Bierme, P., Colombies, J., Ducos, J., Marty, Y. and Ohayon, E.: Nouv. Rev. France, Hematol. *5*, 501 (1965)
134. Ropartz, C., Audran, R., Rivat, L., Rousseau, P. Y. and Fine, J. M.: Vox Sang. *8*, 627 (1963)
135. Kolins, J., Holland, P. and McGinness, M.: Cancer *42*, 2248 (1978)
136. Hocking, D.: Med. J. Aust. *2*, 402 (1971)
137. Salmon, C.: Blood Cells *2*, 211 (1976)
138. Kassulke, J., Hallgren, H. and Yunis, E.: Am. J. Pathol. *56*, 533 (1969)
139. Kassulke, J. et al.: J. Natl. Cancer Inst. *46*, 1201 (1971)
140. Schenkel-Brunner, H., Chester, M. and Watkins, W.: Eur. J. Biochem. *30*, 269 (1972)
141. Kuhns, W., Oliver, R., Watkins, W. and Greenwell, P.: Cancer Res. *40*, 268 (1980)
142. Atkinson, J., Tanley, P. and Wallas, C.: Transfusion *23*, 408 (1983)
143. Testa, U., Henri, A., Bettajeb, A. et al.: Cancer Res. *42*, 4694 (1982)
144. Kannagi, R., Pappayanopoulou, T., Nakamoto, B. et al.: Blood *62*, 1230 (1983)
145. Shumak, K., Beldotti, L., Rachkewich, R.: Brit. J. Haematol. *41*, 399 (1979)
146. Bird, G., Wingham, J., Pippard, M., Hoult, J. and Melikian, V.: Brit. J. Hem. *33*, 289 (1976)
146a. Ness, P., Garratty, G., Moral, P. and Perkins, H.: Blood *54*, 30 (1979)
147. Baldwin, M., Barrasso, C., Ridolfi, R.: Am. J. Clin. Path. *72*, 1024 (1979)

148. Hoogstraten, B., Rosenfield, R. and Wasserman, L.: Transfusion *1,* 32 (1961)
149. Kahn, A., Vroclans, M., Hakim, J. and Biovin, P.: Lancet *2,* 933 (1971)
150. Whitkamp, L., Sing, C., Schreffler, D. and Guttormsen, S.: Am. J. Hum. Genet. *21,* 600 (1969)
151. Salmon, C., Andre, R. and Philippon, J.: Rev. Fr. d'Etude Clin. Biol. *6,* 792 (1961)
152. Majsky, A.: Folia Haematol. *87,* 249 (1961)
153. Salmon, C. and Carton, J.: ABH Blood Group Changes, in: Handbook Series in Clinical Laboratory Science (Seligson, J. (Ed.)), CRC Press, Boca Raton, 1977
154. Starling, K. and Fernbach, D.: Transfusion *10,* 3 (1970)
155. Salmon, C., Cartron, J. and Lopez, M.: Leuk. Res., Manuscript submitted
156. Salmon, C., Cartron, J. and Rouger, P.: Blood Groups and Pathology: An Introductory Survey in Blood Groups and other Red Cell Surface Markers in Health and Disease (Salmon, C. (Ed.)), Masson, New York, 1983
157. Crookston, M.: AB0, H and Ii Phenotypes, in: Disease: in Blood Groups in Disease (Garratty, G. (Ed.)), Amer. Assoc. Blood Banks Monograph, 1983
158. Lehrs, H.: Z. Immunitätsforsch. *66,* 175 (1930)
159. Putkonen, T.: Acta Soc. Med. Fenn., "Duodecim", A, *14,* No. 12 (1930)
160. Schiff, F. and Sasaki, H.: Klin. Wochenschr. *11,* 1426 (1932)
161. Friedenreich, V. and Hartmann, G.: Z. Immunitätsforsch. Immunobiol. *92,* 141 (1938)
162. Wiener, A. S. and Kosofsky, I.: J. Immunol. *42,* 381 (1941)
163. Ohuti, K. and Tohoku, J.: Exp. Med. *51,* 297 (1949)
164. Masamune, H., Yosizawa, Z. and Masukawa, A.: Tohoku J. Exp. Med. *58,* 381 (1953)
165. Kay, H. E. and Wallace, B. M.: J. Natl. Cancer Inst. *26,* 1349 (1961)
166. Kawasaki, H. and Tohoku, J.: Exp. Med. *68,* 119 (1958)
167. Davidsohn, I., Kovarik, S. and Lee, C. L.: Arch. Pathol. *81,* 381 (1966)
168. Davidsohn, I., Kovarik, S. and Ni, L. Y.: Arch. Pathol. *87,* 306 (1969)
169. Dabelsteen, E. and Fulling, H.: Scand. J. Dent. Res. *79,* 387 (1971)
170. Hakomori, S., Koscielak, J., Bloch, K. J. and Jeanloz, R. W.: J. Immunol. *98,* 31 (1967)
171. Kay, H. E.: Brit. J. Cancer *11,* 409 (1957)
172. Eklund, A., Gullbring, B. and Lagerloff, B.: Acta Pathol. Scand. *59,* 447 (1963)
173. Davidsohn, I. and Stejskal, R.: Hematologia *6,* 177 (1972)
174. Bergman, S. and Javadpour, N.: J. Urol. *119,* 49
175. Bonfiglio, T. A. and Feinberg, M. R.: Arch. Pathol. Lab. Med. *100,* 307 (1976)
176. Strauchen, J., Bergman, S. and Hanson, T.: Cancer *45,* 2149 (1980)
177. Springer, G., Desai, P., Fry, W., Goodale, R., Shearen, J. and Scanlon, E.: Carcinoma Cell Membrane T Antigen, in: Immunodiagnosis in Membranes in Tumor Growth (Galcotti, T. et al. (Eds.)), Elsevier, New York, 1982
178. Davidsohn, I.: Am. J. Clinical Pathol. *57,* 715 (1972)
179. Sheahan, D., Horowitz, S. and Zamcheck, N.: Am. J. Dig. Dis. *16,* 961 (1971)
180. Kim, Y., Isaacs, R. and Perdomo, J.: Proc. Natl. Acad. Sci. USA *71,* 4869 (1974)
181. Finan, P., Wight, D., Lennox, E. et al.: J. Natl. Cancer Inst. *70,* 679 (1983)
182. Dabelsteen, E., Vedtofte, P. et al.: J. Invest. Dermatol. *9*, 3 (1982)
183. Kapadia, A., Feizi, T., Jewell, D., Keeling, J. and Slavin, G. J.: Clin. Path. *34,* 320 (1981)
184. Schoentag, R., Williams, V. and Kuhns, W.: Cancer *53,* 503 (1984)
185. Denk, H., Tappeiner, G. and Holzner, J.: Europ. J. Cancer *10,* 487 (1974)
186. Cooper, H. and Haesler, W.: Am. J. Clin. Path. *69,* 594 (1978)
187. Cooper, H., Cox, J. and Patchefsky, A.: Am. J. Clin. Path. *73,* 345 (1980)
188. Hakkinen, I.: Nat. Cancer Inst. *44,* 1183 (1970)
189. Paul, A., Hermelin, B., Mergey, M. and Picard, J.: Carbohydrate Res. *110,* 89 (1982)
190. Runge, R. and Pour, P.: Cancer Lett. *10,* 35 (1980)
191. Yokota, M., Warner, G. and Hakomori, S.: Cancer Res. *41,* 4185 (1982)
192. Yoshida, A., Fujii, H., Dave, V., Cozaut, J. and Morel, P.: Blood *56,* 881 (1980)
193. Watkins, W., Greenwell, P. and Yates, A.: Immun. Comm. *10,* 83 (1981)
194. Zweibaum, A., Robin-Leon, S., Simon-Assmann, P., Triadou, N. et al.: Proc. Amer. Assn. Cancer Res. *24,* 153 (1983)
195. Kelly, J. and Alpers, D.: J. Biol. Chem. *248,* 8216 (1973)
196. Decenzo, J., Howard, R. and Irish, C.: J. Urol. *114,* 874 (1975)

197. Flanigan, R., King, C., Clark, T., Cash, J., Greenfield, B., Snieunski, I. and Primus, F. J.: J. Urol. *130,* 499 (1983)
198. Herlyn, M., Steplewski, Z., Herlyn, D. and Koprowski, H.: Proc. Nat. Acad. Sci. USA *78,* 1438 (1979)
199. Koprowski, H., Blaszczyk, M., Steplewski, Z., Brockhaus, M., Magnaui, J. and Ginsburg, V.: Lancet *1,* 1332 (1982)
200. Koprowski, H., Herlyn, M., Steplewski, Z. and Sears, H.: Science *212,* 53 (1981)
201. Magnaui, J., Brockhaus, M., Smith, D., Ginsburg, V., Blaszczyk, M., Mitchell, K., Steplewski, Z. and Koprowski, H.: Science *212,* 55 (1981)
202. Primus, F. J., Kuhns, W., Goldenberg, D. et al.: Proc. Amer. Assoc. Cancer Res. *24,* 203 (1983)
203. Hakomori, S., Nudelman, E., Kannagi, R. and Levery, S.: Biochem. Res. Comm. *109,* 36 (1982)
204. Abe, K., Hakomori, S., Hirata, I. and Ohshiba, S.: Int. J. Cancer; Manuscript submitted
205. Levine, P., Bobbitt, O. B., Waller, R. K. and Kuhmichel, A. B.: Proc. Soc. Exp. Biol. Med. *77,* 403 (1951)
206. Levine, P.: Blood Group Determinants in Malignancy Foreign to Host, in: Onco-Developmental Gene Expression (Fishman, W. and Sell, S. (Eds.)), Academic Press, New York, 1976
207. Breiner, M., Cedergren, B., Karlsson, K., Nilsson, K. and Samuelsson, B.: FEBS Lett. *118,* 209 (1980)
208. Fellous, M., Gerbal, A., Kamoun, M., Chereau, C., Vullien, A. and Dumont, J.: Studies on the P and AB0 Red Blood Cell Systems Using Somatic Cell Genetics, in: Human Blood Groups (Mohn, J., Plunkett, R., Cunningham, R. and Lambert, R. (Eds.)), S. Karger, Basel 1977
209. Levine, P.: Illegitimate Blood Group Antigens P1, A, and M, N (T) in malignancy, A Possible Therapeutic Approach with anti-Tj^a, anti-A and anti-T, in: Immunobiology of Cancer (Southem, C. and Friedman, H. (Eds.)), Ann. N.Y. Acad. Sci. *22,* 428–435 (1976)
210. Kabat, E.: Am. J. Clin. Path. *78,* 281 (1982)
211. Kabat, E., Liao, J., Shyong, J. and Osserman, E.: J. Immunol. *128,* 540 (1982)
212. Burnet, F. and Anderson, S.: Austral. J. Exp. Biol. Med. Sci. *25,* 213 (1947)
213. Springer, G., p. 157, in: Non-HLA Antigens in Health, Ageing and Malignancy (Cohen, E. and Singal, D. (Eds.)), Alan Liss, New York, 1983
214. Springer, G.: Surv. Synth. Path. Res. *2,* 141 (1983)
215. Springer, G., Murthy, S., Desai, P., Fry, W., Tegtmeyer, H. and Scanlon, E.: Klin. Wchschr. *60,* 121 (1982)
216. Springer, P.: Naturwissenschaften *70,* 621 (1983)
217. Springer, G.: Personal communication
218. Cooper, H.: Lab. Invest. *47,* 383 (1982)
219. Boland, C., Montgomery, G. and Kim, Y.: Proc. Nat. Acad. Sci. *79,* 2051 (1982)
220. Coon, J., Weinstein, R. and Summers, J.: Am. J. Clin. Path. *77,* 692 (1982)
221. Springer, G., Desai, P. and Scanlon, E.: Cancer *37,* 169 (1976)
222. Milner, J. and Metcalfe, S.: Lancet *2,* 1100 (1982)
223. Samter, U. and Glick, M.: Cancer Res. *43,* 4159 (1983)
224. Krusius, T. and Finne, J.: Eur. J. Biochem. *78,* 369 (1977)
225. Berry, C. and Amerigo, J.: Virchows Arch. A. *388,* 167 (1980)
226. Vedtofte, P. and Dabelsteen, E.: Scand. J. Dent. Res. *89,* 424 (1981)
227. Scott, G. and Rasbirdge, R.: Vox Sang. *23,* 458 (1972)
228. Day. N., Winfield, J., Gee, T., Winchester, R. et al.: Clin. Exp. Immunol. *26,* 189 (1976)
229. Branders, W., Helson, L., Wang, Y., Good, R. and Day, N.: J. Clin. Invest. *62,* 120 (1978)
230. Young, W., Hakomori, S. and Levine, P.: J. Immunol. *123,* 92 (1978)
231. Buchbinder, L.: Arch. Pathol. *19,* 841 (1935)
232. Richards, V.: Progr. Exp. Tumor Res. *25,* 1 (1980)
233. Benjamin, E., Rennich, D. and Sell, S.: Tumor Immunology, Chapt. 17, p. 233; in: Basic and Clinical Immunology (4th Ed.) (Stites, D., Stobo, J., Fudenberg, H., Wells, J. (Eds.)), Lange, Palo Alto, 1982
234. Hamilton-Fairley, G. and Akers, R. J.: Brit. J. Haematol. *8,* 375 (1962)
235. Fernandez, L., MacSween, J. and Langley, G.: Blood *62,* 767 (1983)

236. Lawson, H. A., Stuart, C. A., Paul, A. M., Phillips, A. M. and Phillips, R. W.: New Engl. J. Med. *252,* 13 (1955)
237. Mollison, P. L.: 1971 Medical Res. Council, Spec. Rep. Ser. 310, London, p. 86, 1971
238. Teitelbaum, J. I., Wiener, J. and Desforges, J. F.: J. Lab. Clin. Med. *53,* 535 (1959)
239. Wiener, A. S., Briggs, D. K., Weiner, L. and Burnett, L.: J. Lab. Clin. Med. *51,* 539 (1958)
240. Filitti-Wurmser, S., Jacquot-Armand, U., Aubel-Lesure, B. and Wurmser, R.: Ann. Eugen. (London) *18,* 183 (1954)
241. Filitti-Wurmser, S., Jacquot-Armand, Y. and Wurmser, R.: J. Chim. Phys. *47,* 419 (1950)
242. Filitti-Wurmser, S. and Wurmser, R.: Les etudes quantitatives de l'isohemagglutination, Rapport au X[e] Congr. Soc. Europeene Hematologie, Strasbourg, 1965
243. Salmon, C. and Salmon, D.: Nouv. Rev. Fr. Hematol. *3,* 653 (1963)
244. Ogata, H. and Hasegawa, S.: Transfusion *17,* 651 (1977)
245. Springer, G. F. and Desai, P. R.: Transfusion *17,* 651 (1977)
246. Luner, S. J., Wile, A. G. and Sparks, F. C.: Ann. Meet. Am. Assn. Cancer Res. *375* (Abstr.) 1977
247. Bray, J., MacLean, G., Dusel, F. and McPherson, T.: Clin. Exp. Immunol. *47,* 176 (1982)
248. Springer, G., Murthy, S. and Desai, P.: Klin. Wschr. *60,* 121 (1982)
249. Rouger, P., Riveau, D., Salmon, C. and Loygue, J.: J. Clin. Path. *32,* 907 (1979)
250. Boccardi, V., Attina, D. and Girelli, G.: Vox Sang. *27,* 268 (1982)
251. Dube, V., Iwaki, Y., Anderson, B. and Terasaki, P. O.: Vox Sang. *43,* 113 (1982)
252. Kitimura, H., Levine, P., Cheng, P., Eyeli, R. et al.: Cancer Res. *39,* 2909 (1979)
253. Kijimoto-Ochiai, S., Takahashi, W. and Makita, A.: Japan J. Exp. Med. *51,* 149 (1981)
254. Deegan, M., Hayashi, H. and Sawdyk, M.: Am. J. Clin. Path. *80,* 69 (1983)
255. Knuth, A., Lloyd, K., Lipkin, M., Oettgen, H. and Old, L.: Int. J. Cancer *32,* 199 (1983)
256. Pompecki, R., Shively, J. and Todd, C.: Cancer Res. *41,* 1910 (1981)
257. Bird, G. W. G.: Blut *35,* 165 (1977)
258. Dacie, J. V.: The Haemolytic Anaemias, Part III, Grune and Stratton, New York (1967)
259. Hennemann, H. H. and Beck, T.: Dtsch. Med. Wochenschr. *99,* 1869 (1974)
260. Schreiber, Z. A., Haim, S., Gellei, B. and Tatarsky, I.: Cancer *32,* 922 (1973)
261. Marsh, W. L.: Brit. J. Haematol. *7,* 200 (1961)
262. Rochant, H., Tonthat, H., Etievant, M. F., Intrator, L., Sylvestre, R. and Dreyfus, B.: Vox Sang. *22,* 45 (1972)
263. Garratty, G., Petz, L. D., Wallerstein, R. O. and Fudenberg, H. H.: Transfusion *14,* 226 (1974)
264. Boughton, B. J.: J. Clin. Path. *32,* 523 (1979)
265. Castella, A., LaBarge, B., Lavenstein, K. and Davey, F.: Transfusion *23,* 339 (1983)
266. Herrmann, F., Sieber, G., Komisehke, B., Oestreich, R. and Ruhl, H.: Acta Hemat. *71,* 1 (1984)
267. Springer, G., Murthy, S., Desai, P., Wurtz, K. et al.: Breast *7,* 24 (1982)
268. Hakomori, S. and Murakami, W.: Proc. Natl. Acad. Sci. USA *59,* 254 (1968)
269. Hakomori, S.: Biochim. Biophys. Acta *417,* 55 (1975)
270. Black, P., Hakomori, S. and Warren, L.: Cancer Res. *43,* 2322 (1983)
271. Sundsmo, J. and Hakomori, S.: Biochem. Biophys. Res. Comm. *68,* 799 (1976)
272. Marcus, D. and Schwarting, G.: Adv. Immunol. *23,* 203 (1976)
273. Feizi, T., Turberville, C. and Westwood, J. H.: Lancet *1,* 391 (1975)
274. Wallach, D. F. H.: Membrane Molecular Biology of Neoplastic Cells, Elsevier, Amsterdam, 1975
275. Watanabe, K., Matsubara, T. and Hakomori, S.: J. Biol. Chem. *251,* 2385 (1976)
276. Kannagi, R., Nudelman, E. and Hakomori, S.: Proc. Natl. Acad. Sci. USA *79,* 3470 (1982)
277. Sweeley, C. and Siddiqui, B.: Chemistry of Mammalian Glycolipids, p. 459, in: The Glycoconjugates, Vol. III (Horowitz, M. (Ed.)), Academic Press, New York, 1982
278. Ginsburg, V.: Meth. Enzymol., Vol. 50, Chapt. 60–63, Academic Press, New York, 1978
279. Mora, P., Brady, R., Bradley, R. and McFarland, V.: Proc. Natl. Acad. Sci. USA *63,* 12,909 (1969)
280. Kannagi, R., Levine, P., Watanabe, K. and Hakomori, S.: Cancer Res. *42,* 5249 (1982)
281. Abe, K., McKibbin, J. and Hakomori, S.: J. Biol. Chem. *258,* 11,793 (1983)
282. Magnani, J., Nilsson, B., Brockhaus, M., Zopf, D. et al.: Fed. Proc. *41,* 898 (1982)

283. Brockhaus, M., Magnani, J., Blaszczyk, M., Steplewski, Z. et al.: J. Biol. Chem. *256,* 13,223 (1981)
284. Santer, U., Glick, M., Van Holbuk, H. and Vlieguthurt: Carbohydrate Res. *118,* in press (1983)
285. Cowan, W.: Brit. J. Cancer *16,* 535 (1962)
286. Fellous, M., Covillin, P., Neauport-Sautes, G. et al.: Eur. J. Immunol. *3,* 543 (1973)
287. Cohen, E., Minowada, J., Pliss, M., Pliss, L. and Blumenson, L.: Vox Sang. *31,* 117 (1976)
288. Springer, G., Desai, P., Fry, W., Goodale, R., Shearen, J. and Scanlon, E., in: Cellular Oncology, p. 99,: Cancer Res. Monogr., Vol. 1 (Moloy, P. and Nicolson, G. (Eds.)), Praeger, New York, 1983
289. Piller, F., Cartron, J. and Tuppy, H.: Rev. Fr. Transfus. Immunohematol. *23,* 599 (1980)
290. Holmes, E. and Hakomori, S.: J. Biol. Hem. *258,* 3706 (1983)
291. Matenkura, Y.: Immunology *31,* 571 (1976)
292. Munro, J. and Shachter, H.: Arch. Biochem. Biophys. *156,* 534 (1973)
293. Bernacki, R. and Kim, U.: Science *195,* 577 (1977)
294. Shohet, S. and Kluck, J.: Red Cell Glycolipids and Phospholipids; Composition and Metabolism, in: Membrane Structure and Function of Human Blood Cells (Vyas, G. (Ed.)), Amer. Assoc. Blood Banks, Washington, 1976
295. Weiser, M., Klohs, W., Podolsky, D. and Wilson, J.: Glycosyltransferases in Cancer, p. 301, in: The Glycoconjugates, Vol. 4 (Horowitz, M. (Ed.)), Academic Press, New York, 1982
296. Mulet, C., Cartron, J., Badet, J. and Salmon, C.: FEBS Lett. *84,* 74 (1977)
297. Khilanani, P., Chase, T., Lomen, P. and Kessel, D.: Cancer Res. *37,* 2557 (1977)
298. Kessel, D., Shah-Reddy, I., Mirchandani, I., Khilanani, P. and Chou, T. H.: Cancer Res. *40,* 3576 (1980)
299. Warren, L., Fuhrer, J. P. and Buck, C. A.: Proc. Natl. Acad. Sci. USA *69,* 1838 (1972)
300. Bosmann, H. B. and Hilf, R.: FEBS Lett. *44,* 313 (1974)
301. Henderson, M. and Kessel, D.: Cancer *39,* 1129 (1977)
302. Kloppel, T. M., Keenan, T. W., Freeman, M. J. and Morre, D. J.: Proc. Natl. Acad. Sci. USA *74,* 3011 (1977)
303. Revesz, T. and Greaves, M., 1975, in: Membrane Receptors and Antigens: Relation to Lymphocyte Activation and Differentiation, Internat. Symp. Membrane Receptors of Lymphocytes (Seligmann, M., Prend'homme, J. and Kourilsky, I. (Eds.)), pp. 403–441, ASP and Medical Press, Amsterdam
304. Anstee, D. and Tanner, M.: Biochem. J. *175,* 149 (1978)
305. Springer, G. F. and Yang, H. J.: Immunochem. *14,* 497 (1977)
306. Kuhns, W. J.: Transfusion *18,* 389 (1978)
307. Schenkel-Brunner, H., Prohaska, R. and Tuppy, H.: Eur. J. Biochem. *56,* 591 (1975)
308. Cabezas, J., Reglero, A. and Calvo, P.: Int. J. Biochem. *15,* 243 (1983)
309. Flowers, H. and Sharon, N.: Adv. Enzymol. *48,* 29 (1979)
310. Bosmann, H. B. and Hall, T.: Proc. Nat. Acad. Sci. USA *71,* 1833 (1974)
311. Gold, P. and Freedman, S.: J. Exp. Med. *121,* 439 (1965)
312. Gold, P. and Freedman, S.: J. Exp. Med. *122,* 467 (1965)
313. Rogers, G.: Biochim. Biophys. Acta *458,* 355 (1976)
314. Lo Gerfo, P., Krupey, J. and Hansen, H.: New Engl. J. Med. *285,* 138 (1973)
315. Hansen, H., Primus, F. J., MacDonald, R. and LaFontaine, G.: Fed. Proc. *35,* 275 (1976)
316. Concensus in Medicine: Carcinoembryonic Antigen, its Role as a Marker in the Management of Cancer; Brit. Med. J. *282,* 373 (1981)
317. Kuhns, W., Primus, F. J., Goldenberg, D., Bernard, S., Sandler, R and Herbst, C.: Proc. Amer. Soc. Cancer Res. *24,* 203 (1983)
318. Gold, P., Shuster, J. and Freedman, S.: Cancer *42,* 1399 (1978)
319. Goldenberg, D., Kim, E., DeLand, F., Bennett, S. and Primus, F. J.: Cancer *40,* 2984 (1980)
320. Simmons, D. and Perlmann, P.: Cancer Res. *33,* 313 (1973)
321. Chandraskaran, E., Davila, M., Nixon, D., Goldfarb, M. and Mendicino, J.: J. Biol. Chem. *258,* 7213 (1983)
322. Wiley, E., Mendelsohn, G. and Eggleston: J. Lab. Invest. *44,* 507 (1981)
323. Gold, J. and Gold, P.: Cancer Res. *33,* 2821 (1973)
324. Bali, J., Magous, R., Lecou, C. and Mousseron-Canet, M.: Cancer Res. *36,* 2124 (1976)

325. Hakomori, S., Nudelman, E., Levery, S., Solter, D. and Knowles, B.: Biochem. Biophys. Res. Comm. *100,* 1578 (1981)
326. Ormerod, M.: Scand. J. Immunol. (Suppl.) *8,* 433 (1978)
327. Chism, S., Bell, P. and Warner, N.: J. Immunol. Methods *13,* 83 (1976)
328. Chu, T., Holyoke, E. and Murphy, G.: Cancer Res. *34,* 212 (1974)
329. Wiatzer, G., Uhlenbruck, G., Steinhausen, G. and Germain, H.: Experientia *34,* 255 (1978)
330. Yang, H. and Hakomori, S.: J. Biol. Chem. *246,* 1192 (1971)
331. Holburn, A., Mach, J., MacDonald, D. and Newlands, M.: Immunology *26,* 831 (1974)
332. Coombs, R. and Bedford, D.: Vox Sang. *5,* 111 (1955)
333. Coons, A. and Kaplan, M.: J. Exp. Med. *91,* 1 (1950)
334. Dimmock, E., Franks, D. and Glauert, A.: J. Cell Sci. *10,* 525 (1972)
335. Hogman, C.: Vox Sang. *4,* 12 (1959)
336. Kelus, A., Gurner, B. and Coombs, R.: Immunol. *2,* 262 (1959)
337. Hogman, C.: Vox Sang. *4,* 329 (1959)
338. Hagiwara, A.: Exp. Cell Res. *28,* 615 (1962)
339. Chessin, L., Bramson, S., Kuhns, W. and Hirschhorn, K.: Blood *25,* 944 (1965)
340. Franks, D.: Biol. Rev. *43,* 17 (1968)
341. Friedhoff, F. and Kuhns, W.: Transfusion *8,* 244 (1968)
342. Franks, D. (1967): Antigenic Heterogeneity in Cultures of Mammalian cells, in: Phenotypic Expression: Immunological, Biochemical and Morphological Aspects (Daws, C. J. (Ed.)), Waverly Press, Baltimore
343. Hogman, C.: Exp. Cell Res. *21,* 137 (1960)
344. Fogel, M. and Sachs, L.: Exp. Cell Res. *34,* 448 (1964)
345. Jones, S., Quinn, D., Ma, J., Ward, H., Pihl, E. and Nairn, R.: Pathol. *14,* 405 (1982)
346. Pann, C. and Kuhns, W.: Nature (New Biol.) *240,* 22 (1972)
347. Katoh, Y., Stoner, G., McIntire, K., Hill, T. Anthony, R. et al.: J. Natl. Cancer Inst. *62,* 1177 (1979)
348. Momota, K., Shimoda, K. and Naiki, M.: J. Biochem. *92,* 2047 (1982)
349. Anglin, J., Lerner, M. and Nordquist, R.: Nature *269,* 254 (1977)
350. Codington, J., Sanford, B. and Jeanloz, R.: Biochemistry *11,* 2559 (1972)
351. Springer, G., Codington, J. and Jeanloz, R. W.: Cancer Inst. *49,* 1469 (1972)
352. Matsumoto, M. and Taki, T.: Biochem. Biophys. Res. Comm. *71,* 472 (1976)
353. Smith, E. and McKibbin: J. Anal. Biochem. *45,* 608 (1972)
354. Hiramoto, R., Smith, E., Ghanto, V., Shaw, J. and McKibben, J.: J. Immunol. *110,* 1037 (1973)
355. Meyer, K., Smyth, E. and Palmer, J.: J. Biol. Chem. *119,* 73 (1937)
356. Kabat, E.: Blood Group Substances, Academic Press, New York, 1956
357. Kuhns, W., Pann, C. and Bramson, S.: Am. J. Pathol. *69,* 389 (1972)
358. Jones, H., McKusick, V., Harper, P. and Wu, K.: Obstet. Gynecol. *38,* 945 (1971)
359. Pann, C. and Kuhns, W.: Exp. Cell Res. *98,* 73 (1976)
360. Pann, C.: Phenotypic Expression of Blood Group H on HeLa Cells, Ph. D. Thesis, New York University, 1973
361. Kuhns, W.: J. Cell Biol. *88,* 10a (1978)
362. Bosmann, H., Hagopian, A. and Eylar, E.: Arch. Biochem. Biophys. *128,* 470 (1968)
363. Kaufman, R. and Ginsburg, V.: Exp. Cell Res. *50,* 127 (1968)
364. Yurchenko, P., Ceccarini, C. and Atkinson, P.: Labeling Complex Carbohydrates of Animal Cells with Monosaccharides, in: Methods in Enzymology, Vol. L, Complex Carbohydrates, Part C (Ginsburg, V. (Ed.)), Academic Press, New York, 1978
365. Shen, L. and Ginsburg, V.: Arch. Biochem. Biophys. *122,* 474 (1967)
366. Steiner, S., Brennan, P. and Melnick, J.: Nature (New Biol.) *245,* 19 (1973)
367. Steiner, S., Melnick, J., Kit, S. and Somers, K.: Nature *248,* 682 (1974)
368. Saito, T. and Hakomori, S.: J. Lipid Res. *12,* 257 (1971)
369. Warren, L., Critchley, D. and MacPherson, I.: Nature *235,* 275 (1972)
370. Basu, S. and Basu, M.: Expression of Glycophingolipid Glycosyltransferases in Development and Transformation, p. 265, in: The Glycoconjugates, Vol. 3 (Horowitz, M. (Ed.)), Academic Press, New York, 1982
371. Basu, M. and Basu, S.: J. Biol. Chem. *248,* 1700 (1973)

372. Basu, M., Higashi, H., Basu, S. and Evans, G.: Fed. Proc. *39,* 2184 (1980)
373. Simmons, D.: Bacteriol. Rev. *35,* 117 (1971)
374. Montgomery, R.: Heterogeneity of the Carbohydrate Groups of Proteins, in: Glycoproteins (Gottschalk, A. (Ed.)), pp. 518–528, Elsevier, Amsterdam, 1972
375. Watanabe, K. and Hakomori, S.: J. Exp. Med. *144,* 644 (1976)
376. Merritt, W. and Morre, D. J.: Biochem. Biophys. Acta *620,* 261 (1980)
377. Kemp, R.: Diseases of the Stomach, Intestine and Pancreas (Ed. 3), p. 338, Saunders, Philadelphia, 1917
378. Bockus, H.: Gastroenterol. *1,* 775 (1963)
379. Wilson, G. and Miles, A.: Topley and Wilsons Principles of Bacteriology and Immunity, 5th Ed., Williams and Wilkins, Baltimore, 1964
380. Springer, G. and Horton, R.: J. Gen. Physiol. *47,* 1229 (1964)
381. Watkins, W., Kuhns, W., Greenwell, P. and Anderson, A.: Commun. 17th Congr. Internat. Soc. Blood Transfusion 5-41-5, 85 (1982)
382. Clarkson, B.: Survival Value of the Dormant State in Neoplastic and Normal Cell Populations, in: Control of Proliferation in Animal Cells (Clarkson, B. and Baserga, K. (Eds.)), Cold Spring Harbor Laboratory, Cold Spring Harbor, New York, 1974
383. Rowley, J.: Nature *301,* 240 (1983)
384. Inbar, M.: Dynamics of Cell Surface Membranes in Leukemia, in: Cell Membranes Tumor Cell Behavior, pp. 9–18, 28th Ann. Symp. Cancer Res., M. D. Anderson Hospital, Tumor Inst., Houston, Texas, 1975
385. Zabel, P., Noujain, A., Shysh, A. and Bray, J.: Eur. J. Nuc. Med. *8,* 250 (1983)
386. Currie, G.: Cancer and the Immune Response, Arnold, London, 1974
387. Yonezawa, S., Nakamura, T., Tanaka, S. and Sato, E.: J. Natl. Cancer Inst. *69,* 777 (1982)
388. Yonezawa, S., Nakamura, T., Tanaka, S., Maruto, K., Nishi, M. and Sato, E.: J. Natl. Cancer Inst. *71,* 19 (1983)
389. Hansson, G., Karlsson, K., Larson, G., McKibbin, J., Blaszczyk, M., Herlyn, M., Steplewski, Z. and Koprowski, H.: J. Biol. Chem. *258,* 4091 (1983)
390. Watkins, W.: Rev. Fr. Transfus. *21,* 201 (1978)
391. Stellner, K., Hakomori, S. and Warner, G.: Biochem. Biophys. Res. Commun. *55,* 439 (1973)
392. Primus, F. J., Clark, C. and Goldenberg, D.: Immunochemical Detection of Carcinoembryonic Antigen, Chapt. 16, p. 263, in: Diagnostic Immunohistochemistry (DeLellis, R. (Ed.)), Masson, New York, 1981, 1981
393. Giblett, E. and Crookston, M.: Nature *201,* 1138 (1964)
394. Hysell, J. K., Gray, J. M. and Beck, M. L.: Transfusion *14,* 72 (1974)
395. De Bruyere, M., Sokal, G., Devoitille, J. M., Fauchet-Dutrieux, M. Ch. and de Spa, V.: Brit. J. Haematol. *20,* 83 (1974)
396. Bernstein, D., Near, S., Rikover, M. and Menahem H.: Obstet. Gynecol. *43,* 276 (1974)
397. Gordon, P. A., Baylis, P. H. and Bird, G. W. G.: Brit. Med. J. *1,* 1569 (1976)
398. Bellou, B., Jaffe, R., Taylor, R., Solter, D. and Hakala, T.: J. Immunol. *132,* 2111 (1984)
399. Hakomori, S., Nudelman, E., Levery, S. and Kannagi, R.: J. Biol. Chem. *259,* 4681 (1984)
400. Fukushi, Y., Hakomori, S., Nudelman, E. and Cochran, N.: J. Biol. Chem. *259,* 4681 (1984)
401. Walker, P., Karick, S., Dekernion, J. and Pranberg, J.: Am. J. Clin. Path. *81,* 503 (1984)
402. McGowan, A., Moore, S., Chirnard, A. and Yap, P. L.: Hybridoma *3,* 91 (1984)
403. Kelton, J. G. and Babcock, G.: Clin. Res. *32,* 311A (1984)
404. Greenwell, P., Ball, M. and Watkins, W.: FEBS Lett. *164,* 314 (1983)
405. Hoskins, L. C.: Degradation of Mucus Glycoproteins in the Gastrointestinal tract, in: The Glycoconjugates, Vol. II (Horowitz, M. I. and Ward Pigman (Eds.)), Academic Press, New York, p.21978
406. Koenderman, A., Shiphurst, W. and Van Den Eijnden, D.: Proc. XII Int. Carbohydrate Symp. Utrecht, Netherlands, p. 234, 1984
407. Parker, C., Soldato, C. and Rosse, W.: J. Clin. Invest. *73,* 1130 (1984)
408. Gooi, H., Feizi, T., Kapadia, A. et al.: Nature *292,* 156 (1981)
409. Schachter, H.: Clinical Biochem. *17,* 3 (1984)
410. Landsteiner, K.: Zbl. Bakt. *27,* 357 (1900)
411. Kannagi, R., Levery, S. and Hakomori, S.: FEBS Lett. *175,* 397 (1984)
412. Kumazaki, T. and Yoshida, A.: Proc. Natl. Acad. Sci. USA *81,* 4193 (1984)

Copper and Inflammation

Ulrich Deuschle and Ulrich Weser

Anorganische Biochemie, Physiologisch-chemisches Institut der Universität Tübingen,
Hoppe-Seyler-Str. 1, 7400 Tübingen 1, F.R.G.

Copper belongs to the biochemically active and essential transition metals. At present more than twenty different copper proteins are known. Their molecular properties and functional side is well characterized. In each of these proteins the copper is coordinated to the respective polypeptide side chains and is, in fact, the active centre. Copper exerts several distinct properties including the positive redox potential, the pronounced chelating activity and the extraordinary kinetic properties. Attributable to these properties this transition metal is preferentially active in electron transport reactions. Oxygen belongs to the preferred substrates. Very little is known on the biochemical reactivity of copper in inflammation. One of the major causes of inflammation is assigned to the action of excited oxygen species. Quite frequently, a rise of the copper concentration in serum up to a factor of four is observed in the course of inflammatory diseases. Extraneously administered copper has an antiinflammatory effect. We do not know whether or not the reactivity of copper must be exclusively seen in the biochemistry of oxygen. It is attempted to summarize the present thoughts and facts on the general biochemistry of both copper and inflammation. It will be attempted to correlate the biochemical properties of this most prominent transition metal with the biological aspects of inflammation.

Progress in Clinical Biochemistry
and Medicine, Vol. 2

1 Introduction

In contrast to the present age of antibiotics and synthetically prepared pharmaceuticals, copper-containing drugs enjoyed wide application for treatment of many diseases in the past. Throughout the written history of mankind there have been many recipes dealing with copper, not only as a trace element. A medical encyclopaedia originating from the Middle Kingdom of Ancient Egypt (approximately 1550 B.C.) known as the "Papyros Ebers"[1] (Fig. 1) is a promiment example. Suspensions or ointments containing copper hammering flakes or malachite, together with other components, were used for treatment of pain in the joints. More than twenty different recipes using copper minerals, slag derived from copper mining, verdigris as well as elemental copper are known. They were applied both internally and externally. In his encyclopaedia "Naturalis Historia"[2] they are summarized by C. P. Plinius the Elder (23–79). These remedies were reported to heal a great number of diseases of the skin as well as of the eyes. They were also beneficial toward healing ulcers, wounds of the skin and inflamed mucous membranes. At the beginning of the 19th century, Rademacher reexamined the studies of Paracelsus (1493–1541). He deduced, after a 25-year clinical trial, that the body has a strange predisposition to a great number of diseases which disappeared, or could be healed, in the presence of copper. Healthy subjects did not show any responses to copper[3]. At the end of the 19th century, copper was administered internally and found effective against intestinal complaints, diseases of the skin, infections (tuberculosis, syphilis). Certain recovery from neurological disorders, including chorea and neuralgia, was also seen following the administration of copper[4]. At the beginning of this century, many non-copper drugs were developed which were thought to be more effective than previous ones. Thus, the use of copper in general medicine was progressively diminished.

However, the significance of copper in general biological sciences was unbroken. In 1928 it was discovered that copper belongs to the essential trace elements in animals[5]. Furthermore, many physiological functions proved to be dependent on the presence of copper. A striking phenomenon was the dramatically increased concentration of serum copper concomitant with inflammation. This observation has to stand up to rigorous questioning. What is the advantage of the organism when mobilizing copper? It is attractive to conlcude that copper plays an important role in the biological defence and repair mechanisms that together comprise the phenomenon of inflammation. This proposal is supported by the inhibitory activity of administered copper on the development of an inflammatory process. Ironically, this correlation of an increased serum copper level and the intensity of an inflammation was initially seen the other way round. High copper levels were thought to be the actual cause of the inflammatory process.

What is the significance of copper in the general course of an inflammation? What is the mechanism of the copper-dependent inhibition? It will be attempted to summarize both the biochemistry and pharmacology of copper in the mammalian organism. Emphasis will be placed on the biochemical aspects of inflammation. Our limited knowledge in these three large sections should be focused on trying to reveal the possible mechanism of the anti-inflammatory activity of copper.

Fig. 1. Papyros of Ebers in hieratic script. Two examples of recipes which contain copper: LXXIX: Another to expel pain (?) in all limbs of a man: ammi, styrax, honey, precious (?) malachite, dog's skin are applied to his face, thou shalt not lay hands on him. Another to soften the member of the knee: northern salt, gum ammoniac, grease of ibex, honey, frankincense, celery, *ḥ*nj-t3, calamine (?), onion, hammering-flakes from copper, grease of sheep, cumin, oil, natron are ground and (it) is bandaged therewith

Table 1. Exchange rate of coordinated water in the axial position of metal aquo complexes[6] $[M(H_2O)_6] + H_2O^{18} \rightarrow [M(H_2O)_5(H_2O^{18})] + H_2O$

aquo complex	k_1 (s^{-1})
Cr^{3+}	5×10^{-7}
Al^{3+}	10^0
Fe^{3+}	3×10^3
Mg^{2+}	$> 10^4$
Ni^{2+}	3×10^4
Co^{2+}	10^6
Fe^{2+}	3×10^6
Mn^{2+}	3×10^7
Cu^{2+}	8×10^9

2 The Biochemistry of Copper

Copper belongs to the biochemically active and essential elements. It is certainly not a fortunate happenstance that copper plays a key role in biochemical systems. There are many distinct criteria which support its chemical reactivity. Copper is an extremely good chelating cation and the redox potential is on the positive side. Copper(II) is a most prominent example for the well-known Jahn-Teller phenomenon. Cu 3 d^9 is unstable in highly symmetric tetrahedral and octahedral ligand fields. Thus, it is usually found in distorted tetragonal environments. Quite frequently the axial ligand is further apart which results in a kinetically rather unstable situation. By way of contrast, the chromium(III)-aquo complex belongs to the kinetically most stable complexes. In fact, the exchange rate of the bivalent copper-aquo complex, compared to the chromium-aquo complex, proceeds by 17 orders of magnitude faster (Table 1). The positive redox potential explains the active participation in biological electron transport reactions (see Table 2). The electron-accepting oxygen is very loosely bound. For example, the electron transport of four electrons on to molecular oxygen proceeds with utmost velocity in the case of cytochrome oxidase. Due to this fast reaction, possible intermediary products such as superoxide or peroxide are only hypothesized. Unfortunately, there is no experimental technique available to allow the measurement of this fast reduction of oxygen. At the same time, the structure and function correlation of Cu_2Zn_2superoxide dismutase should be emphasized (Fig. 2). This binuclear Cu-Zn-protein reacts with superoxide radicals at a rate which belongs to the fastest reactions ever measured in biochemical systems. The pseudo first-order rate constant is approximately $2 \cdot 10^9\ M^{-1}\ s^{-1}$ and is practically diffusion-limited.

The dominant role of copper in the biochemistry of oxygen should also be seen under the above-mentioned criteria. There is an important evolutionary aspect in that oxygen accumulated in the primordial atmosphere and anaerobic life had to adapt to the aerobic environment. In the course of this process, specific transition metals (i.e. copper, iron) proved particularly suitable to form the active centres of

Table 2. Functional aspects of some copper proteins

Copper protein	Activity	Function	Lit.
haemocyanin	reversible oxygenation	respiratory protein in molluscs	7
Cu_2Zn_2superoxide dismutase	catalytical dismutation of superoxide	copper transport protein and/or removal of superoxide (intracellular)	8–11
lysyl oxidase	oxidative linkage of the ε-amino groups of lysyl residues in collagen and elastin	cross linkage (maturation) of procollagen and proelastin	12, 13
monoamine oxidase	oxidative desamination of amines	catabolism of biogenic amines (i.e. adrenergic neurotransmitters)	14
diamine oxidase	oxidative desamination of diamines	catabolism of diamines (also histamine)	15
peptidyl-α-amidating activity	formation of an amide $-XNH_2$ from a peptide sequence –X–Gly–Y– (cosubstrate ascorbate); Gly reacts to glyoxylate	biosynthesis of peptide neurotransmitters and hypophysis factors with amidated function	16
tyrosine hydroxylase	o-hydroxylation of tyrosine, oxidation to the o-quinone	biosynthesis of dopaminergic neurotransmitters, melamin formation	17
dopamine-β-monooxygenase	β-hydroxylation of dopamine	biosynthesis of adrenergic neurotransmitters	17, 18
cytochrome oxidase	electron transport from cytochrome c to O_2 in the respiratory chain	energetic coupling of the oxidative phosphorylation to the electron transport in the respiratory chain, vectorial H^+ transport in the inner mitochondrial membrane	19, 20
coeruloplasmin	stoichiometric scavenging of superoxide and hydroxyl radicals, oxidase (Fe^{2+}, bifunctional aromatic amines and phenols, ascorbate); blue oxidase	iron metabolism, copper transport, antioxidative in serum, acute phase protein (α_2-globulin)	21–23
ascorbate oxidase	blue oxidase, reduction of $O_2(0)$ to $2\,O(-II)$	terminal oxidase in plants	24
laccase	blue oxidase, reduction of $O_2(0)$ to $2\,O(-II)$	oxidation of mono- and diphenols in plants	24
metallothionein	unknown	copper storage, possibly copper donator for the biosynthesis of copper proteins	25–27
azurin plastocyanin stellacyanin	electron transport (blue copper proteins)	electron transport in the respiratory chain of bacteria and plants	24

many essential metalloproteins. By looking at the different copper proteins of Table 2, it is not surprising that molecular oxygen, superoxide, peroxide and even water are metabolized by many of these proteins. Throughout, the copper remains more or less the same; however, the protein ligand dictates the respective specific biochemical reactivity of the chelated copper.

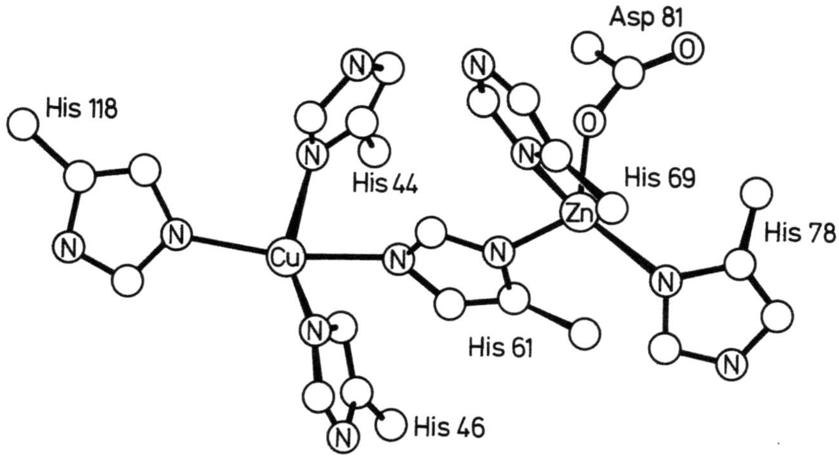

Fig. 2. Active site of bovine Cu_2Zn_2superoxide dismutase. Bovine Cu_2Zn_2superoxide dismutase (M_r approx. 32000) consists of two identical subunits. Their active sites contain one Cu(II) and Zn(II), respectively, which are linked by a common ligand (His 61). Copper is essential for the activity of the enzyme and is known to catalyze the disproportionation of superoxide by cyclic reduction and oxidation, zinc can be absent; in this case the enzymic activity is slightly reduced. The role of zinc is not fully understood; it might be responsible for structural stabilization of the protein conformation. The structure of the active site of Cu_2Zn_2superoxide dismutase has been drawn using the atomic coordinates published by Trainer et al.[28]

Unlike the haemoproteins where the iron is bound to a specific porphyrin residue, copper is always chelated to specific sites of the polypeptide chain of the protein. According to Malkin and Malmström[29], there are three different types of chelated copper. They are characterized by their distinct differences of physico-chemical properties.

Type I Cu^{2+}: This chelated Cu^{2+} exhibits an intensive electron absorption ($\varepsilon_{Cu} \sim 5000$) near 600 nm and intriguing magnetic properties. In the EPR spectrum an unusually small hyperfine coupling constant ($A \sim 0.006\ cm^{-1}$) is detected. In addition, a relatively high reduction potential in the 300–500 mV range vs. 150 mV for Cu(II) is seen[21,29–31]. Usually two histidine ligands, one thiolate sulphur and a

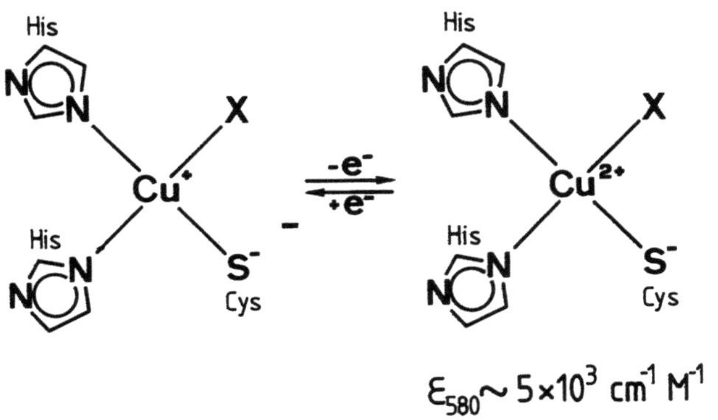

Fig. 3. Oxidation-reduction of the type I or "blue"-copper chromophore

second sulphur ligand (for example, disulphide, methionine or RC=S), contribute to these spectroscopic properties. In the case of the second sulphur ligand, however, non-sulphur ligands cannot be fully excluded (Fig. 3).

There is a general agreement[32] that the strong electronic absorption in the 600 nm region is associated with π S and/or σ S → Cu ($d_{x^2-y^2}$) charge transfer transition. The cysteine ligand is presumed to be the potent electron donor. This was supported by model studies of low energy S (thiolate) → Cu(II) charge transfer transitions using the tetrahedrally constrained environment of 3 d^9 copper bound in 3,5-dimethyl-1-pyrazole borate[33].

Type II Cu^{2+}: This copper is characterized by a weak electronic absorption (ε_{Cu} ~ 50–150). Occasionally it is also termed "non-blue" copper. The EPR properties are similar to those exhibited in distorted tetragonal environments. Unlikely the many low molecular weight Cu(II)-chelates or other "non-blue" copper-proteins including Cu_2Zn_2superoxide dismutase, galactose oxidase and benzylamine oxidase, this type II copper has very unique chemical properties associated with the simultaneous presence of type I and type III copper binding centres found in multi-copper oxidases. Thus, this type II definition should be reserved for the "non-blue" copper binding sites in the blue multi-copper oxidases.

Type III Cu^{2+}: The type III copper binding centre is a binuclear copper complex. It is found in haemocyanin, phenoloxidases and in all multi-copper oxidases. Electronic absorption is measured at 330 nm, no EPR spectrum is detectable with the consequence of non-paramagnetism over a wide range of temperature. The single 3 d^9 electrons are spin-coupled, which explains the observed diamagnetism. The best guess as deduced from many spectroscopic data including EXAFS measurements includes a bridging ligand, probably tyrosine[34–36] (Fig. 4).

As stated above, the protein portion of a copper protein dictates the respective biochemical specificity. The binuclear copper binding centre of haemocyanin is well able to act as a respiratory system in molluscs. Dioxygen is bound to two coppers in the oxy-haemocyanin. The monovalently charged superoxide radical reacts with the "non-blue" copper binding site in the Cu_2Zn_2superoxide dismutase. Changes in the polypeptide chain can cause copper to react with the peroxide anion. The rather sophisticated cytochrome c oxidase is capable of transferring four electrons onto molecular oxygen to produce water in the respiratory chain. Apart from the well characterized copper proteins, there are low molecular weight copper chelates of

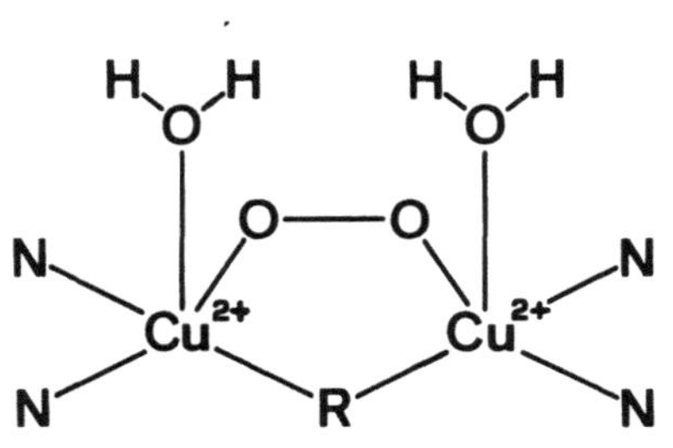

Fig. 4. Binuclear or type III copper binding centre

peptides and amino acids. Attributable to their superoxide dismutase activity, their possible role in scavenging excited oxygen species is discussed. They might also be quite good carriers in copper transport systems, especially in the serum.

3 Copper Metabolism

3.1 Serum Copper

Some aspects of copper metabolism will be discussed below. Mammalian serum contains approximately 15 μM copper (1 μg/ml). Roughly 93% are found coordinated in coeruloplasmin[37–39]. The residual portion is bound to albumin and – to a lesser extent – coordinated in low molecular weight copper chelates. The two major complexes are copper-albumin and a ternary histidine copper serum albumin complex[40]. 0.2–0.4% (2–4 nM) of the non-coeruloplasmin bound copper passed membrane filtration[41] in low molecular weight form (Fig. 5). Likely candidates of these

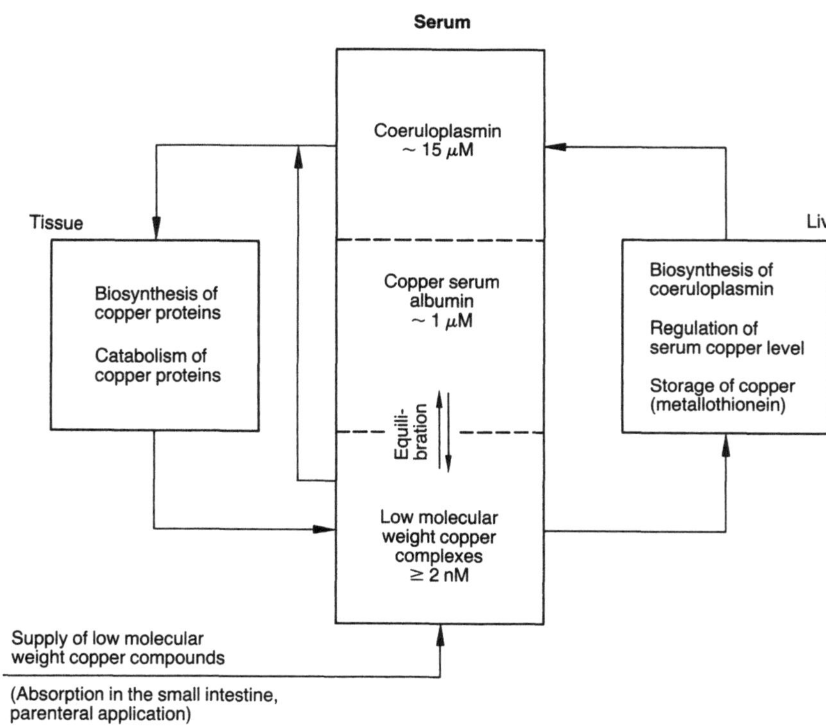

Fig. 5. Copper in the mammalian organism. The concentrations of copper complexes in serum of normal individuals are given. The copper content of the liver depends on both nutrition and state of health. The liver could be assigned to act as a "copper buffer"; it affects the availability of copper in other organs by regulating the serum copper level

copper complexes were copper-histidine and ternary complexes employing histidine and another amino acid (preferentially threonine, cystine[42, 43]). In comparing the binding constants of copper serum albumin and copper-histidine (Table 3), the expected concentration of low molecular weight copper chelates in serum would be approximately 1 pM. At present, we do not know whether this discrepancy may be attributed to erroneous experiments, or whether there is an unknown low molecular weight copper chelating ligand (for example peptide) available.

Under normal physiological conditions, the copper of coeruloplasmin does not exchange with the copper found in the other fractions[37, 38, 49]. As a possible link between the coeruloplasmin and albumin compartments of copper, the catabolism of coeruloplasmin may be seen. By way of contrast, the copper serum albumin complex is kinetically labile. There is a fast equilibrium with the low molecular weight serum copper[42, 43] and with the tissue copper[50]. Furthermore, there is expected to be a fixed ratio between copper serum albumin and low molecular weight copper chelates[43] in the serum. This is explained by a high excess of the ligands over the metal (serum albumin approx. 700 μM, histidine 130 μM, all amino acids totally 40 mM[51, 52]).

A well known phenomenon of the systemic inflammation is the intriguing rise of the serum copper concentration (see also 5.2). All three copper binding systems including coeruloplasmin, albumin and the low molecular weight chelates are substantially higher[39, 53–55]. Furthermore, the hepatic biosynthesis of coeruloplasmin is stimulated[56]. We do not know the origin of the copper incorporated into the other fractions. Accelerated turnover of coeruloplasmin, as usually observed in inflammatory diseases, could be the source[57].

The molecular basis of Cu-transport deserves special attention. The high binding capacity to proteins, its ability to create reactive oxygen species[25, 27] and its essential biochemical role in many oxidases, demand a well-balanced regulation of the cellular copper. As free Cu(II)-ions can undergo many uncontrolled reactions, Cu has to be kept in the stable Cu(I)-thiolate state. Different reaction mechanisms should be discussed in the metabolism of Cu. Firstly, there is a direct transfer of Cu(I) which is facilitated by the hydrophobic nature of the Cu(I)-thiolate chromophore[58, 59]. Sec-

Table 3. Thermodynamic stability of some copper complexes

Ligand	log K_1	log K_2	Lit.
Formate	1.57	0.65	44
Acetate	1.67	0.98	45
Salicylate	10.45	18.47	46
Histidine*	10.2	18.5	47
Serum albumin	16.2	–	48

$$K_1 = \frac{[CuL]}{[Cu][L]} \qquad K_2 = \frac{[CuL_2]}{[Cu][L]^2} \qquad L = \text{Ligand}$$

* Data are given for $[CuHis]^{2+}$ and $[CuHis_2]^{2+}$, respectively; at pH 7.4 the latter species comprises 95% of all possible complexes

ondly, thionein-Cu can provide the organism with essential Cu(II) through a catabolic mechanism, i.e., an enzymically-controlled oxidative breakdown of the Cu(I)-thiolate chromophore by a specific Cu-thionein oxidase.

On the search for an oxidative pathway of Cu(II)-formation, it was attempted to use xanthine oxidase as a possible "Cu(I)-thionein-oxidase"[60]. The Cu(I)-thiolate chromophores of Cu-thionein were oxidized in the presence of this flavoenzyme, as deduced from spectrometrical measurements using EPR and circular dichroism. Unlike Cu_2Zn_2superoxide dismutase, catalase inhibited the oxidative cleavage, suggesting peroxide as the actual oxidizing agent. It might be assumed that there is an enzymically-controlled oxidative pathway for the generation of biologically important Cu(II).

$$Cu(I)\,(^{-}SR)_2 \xrightarrow[-3e]{} Cu(II) + RS\text{-}SR$$

As a spontaneous oxidation of the Cu-thiolate chromophore in vivo does not occur, this "catabolic" metal metabolism would explain the need for a specific Cu-thionein oxidase.

Although there is no proof of a proteolytic cleavage of native copper-thionein, this possible mechanism of Cu(II) formation should also be considered.

It might be attractive to see a regulatory mechanism of generated Cu(II) which is not immediately bound by apo-copper-proteins. This Cu(II) would inhibit xanthine oxidase and bring the further oxidation of Cu(I)-thionein to a halt. Furthermore, the well known iron/copper relationship should be seen in this context. High amounts of liver catalase would be able to scavenge the flavoprotein-dependent generation of H_2O_2, leading again to the inhibition of Cu(II) release from Cu(I)-thionein. Further experiments in vivo will be awaited with great interest.

3.2 Copper Transport Into the Tissue

In serum, the half life of the copper serum albumin complex is roughly 10 min. The decrease of copper in the blood parallels an increase in the liver[61]. In the presence of serum albumin, amino acids are able to stimulate copper transport into liver slices or erythrocytes[62–64]. Likewise, the copper-chelating tripeptide Gly-His-Lys increases copper absorption in hepatoma cells[65]. This effect is not seen when histidine is added to isolated liver cells in the absence of serum albumin[66]. In the presence of serum albumin, low molecular weight ligands are able to sequester only a small part of the copper bound to albumin, which might, however, be sufficient to facilitate copper transport[67].

There is no doubt that coeruloplasmin has a copper transport function[21–23, 68]. Coeruloplasmin copper is incorporated into cytochrome oxidase quite rapidly when copper-deficient rats are used[69]. Lysyloxidase activity in the aorta of copper-deficient chicks is only detectable after the level of coeruloplasmin in the serum has risen[70]. The angiogenesis (formation of blood vessels) in rabbit cornea is stimulated by coeruloplasmin and other copper complexes[71]. The half life of coeruloplasmin in the serum of rats is 13 hours[68]. It is interesting to note that under physiological pH conditions coeruloplasmin copper is only exchangeable when there is an enzymic activity of coeruloplasmin, in other words when there is a substrate present, for example, p-phenylene-diamine[72].

4 Pharmacology of Copper

In the middle of this century, synthetically-prepared copper chelates were successfully used to treat inflammatory diseases, especially those of rheumatism and rheumatoid arthritis[73–75]. In the last twenty years, attempts were made to elucidate their mode of action. The anti-inflammatory activity of many different copper compounds, including inorganic copper salts, amino acid chelates, copper salicylates and acetates, were compared and studied in animal systems (see also 5.3)[76–83].

When the copper complexes were given parenterally, no significant difference in their anti-inflammatory activity was seen. Copper complexes of established anti-inflammatorily active ligands, for example, glucocorticoids and non-steroidal anti-inflammatory drugs, have been shown to be more potent compared to the free ligand. It was hypothesized that these copper chelates were the actual reactive species[76]. Upon oral administration of these copper chelates this anti-inflammatory activity was diminished[77, 81, 82]. It must be assumed that these copper complexes dissociate in the intestinal tract[84] with the consequence of a reduced copper absorption. An anti-inflammatory activity was detected when copper chelates were applied onto the skin[85, 86]. Proof of a penetration was successful using ^{64}copper chelates[85]. Likewise, implanted elemental copper was also anti-inflammatory[87]. In the presence of oxygen a detectable portion of elemental copper is dissolved:

$$2\,Cu^0 + O_2 + 4\,HL \rightarrow 2\,CuL_2 + 2\,H_2O$$

L = ligand (for example amino acid or hydroxy carbonic acid)

Copper intrauterine devices (IUDs)[88] are known to release copper. In addition to the inhibition of embryogenesis including impaired implantation and blastocyst development copper IUDs might have an anti-inflammatory effect which is, however, not yet established[87]. At the same time, a possible spermicidal activity cannot be excluded. Arthritic patients carrying copper bracelets were reported to experience subjective improvement. In the course of the trial a weight loss of the copper bracelets was measured. Milligrams of copper are dissolved daily in sweat[89]. Copper complexes with amino acids (preferentially glycine) are formed and transported through the skin[90].

The copper proteins Cu_2Zn_2superoxide dismutase[91–94] and coeruloplasmin[95] are reported to be anti-inflammatorily active. The general conclusion seems to be warranted that copper which bypasses the regulation of copper absorption controlled by the intestinal mucosa or liver has anti-inflammatory activity. On the other hand, copper deficiency aggravates inflammation[96, 97]. The pharmacological experiments are rather unsatisfactory in that they do not reveal the possible mode of action of copper in inflammation. The animal used in the experiments is considered as a "black box". The compound to be tested is administered and the effect on experimentally induced inflammation is detected. Unfortunately, the molecular action of copper in inflammation remains in the dark.

5 Inflammatory Process

5.1 Local Reaction

There are many different causes which can induce tissue damage resulting in inflammation. Injuries, microbial infections and (auto)-immunoreactions have to be listed under the possible causes. Cells adjacent to the damaged site are activated (Fig. 6):
- mast cells secrete histamin which is stored in granules
- thrombocytes are stimulated by free collagen and secrete serotonin
- kininogenases of serum are activitated in a similar manner
- leukocytes are synthesizing prostaglandins and leukotrienes[98–102]
- phagocytes (polymorphonuclear granulocates, monocytes, tissue macrophages) produce reactive oxygen species[103–106] (Fig. 7); in addition lysosomal enzymes are liberated[109].

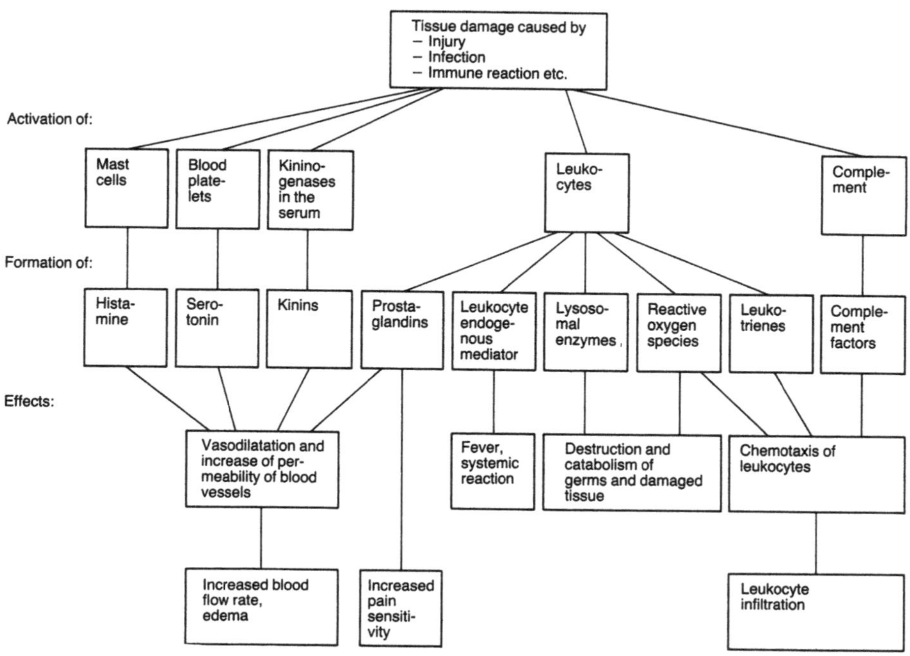

Fig. 6. The local inflammatory reaction. The classical symptoms of inflammation (dolor, rubor, calor, tumor = pain, reddening, warmth, swelling) are essentially explained by vascular effects and leukocyte infiltration

b)

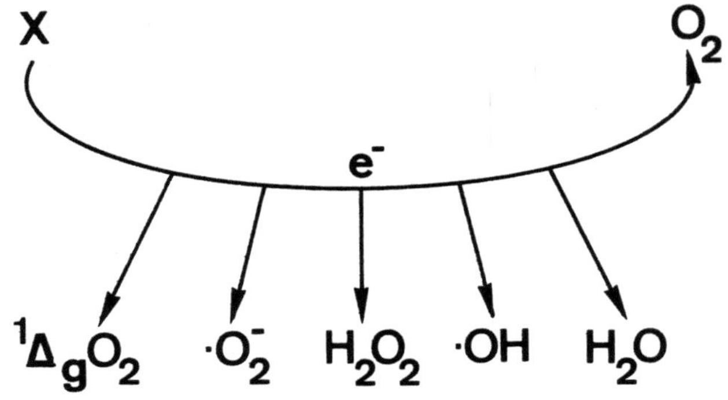

a)

$$M_{aq}^{(n+1)+} + \cdot O_{2\,aq}^{-} \longrightarrow M_{aq}^{n+} + O_2$$

$$M_{aq}^{n+} + H_2O_2 \longrightarrow \cdot OH + OH_{aq}^{-} + M_{aq}^{(n+1)+}$$

$$\cdot OH + \cdot O_{2\,aq}^{-} \longrightarrow O_2 + OH_{aq}^{-}$$

$$H_2O_2 \xrightarrow{h\nu} \cdot OH + \cdot OH$$

$$\cdot OH + RH \longrightarrow R\cdot + H_2O$$

$$R\cdot + O_2 \longrightarrow ROO\cdot$$

$$ROO\cdot + R'H \longrightarrow ROOH + R\cdot'$$

$$3\ {}^1O_2\ ({}^1\Delta_g) \longrightarrow 2\ O_3$$

Fig. 7 a, b. Possible excited oxygen species generated in the course of inflammation and subsequent secondary reactions. **b** Partially reduced oxygen species can be formed during electron transport from a substrate X onto molecular oxygen[107)]. **a** Phagocytes contain NADPH oxidase producing $\cdot O_2^-$ and H_2O_2 on activation[104, 108)]. $^1O_2$, $\cdot OH$ or ferryl species could be subsequently formed. Attributable to their high reactivity biological macromolecules are assumed to be unspecifically attacked. Reactive oxygen species are known to eliminate microbial germs. However, they are also claimed to account for secondary damages induced by the inflammatory process itself

In experimental inflammation induced in animals (i.e. carrageenan paw edema in rats), two phases are distinguishable in the acute reaction[110)]:

- In the first phase histamin, serotonin and kinin induce dilatation and increased permeability of blood vessels. Exudate formation (edema) is the consequence.
- In a later phase these effects are taken over by metabolites of the arachidonic cascade (prostaglandins[110–112)], leukotrienes[101)]). At the same time, reactive oxygen species are discussed to play a vital role[113)]. There is a possible link between

the metabolism of superoxide and the prostaglandin biosynthesis: superoxide is known to stimulate the formation of prostaglandins[91, 113]. More phagocytes are attracted by chemokinetical and chemotactical factors (leukotrienes, complement factors[114], as well as superoxide-induced factors[115–117]).

The local inflammatory reaction effects a modulation of the properties of the blood vessels. Thus, the damaged region is perfectly supplied with the many defensive proteins of the serum, including immunoglobulins, complement factors, acute phase proteins and nutrients. For defence and repair, an intensive metabolism can be maintained. Phagocytes remove damaged cells, microorganisms, complexes between antigens and antibodies and many other irritants. It is likely that the increased levels of anti-inflammatory substances in serum, such as coeruloplasmin and copper amino acid complexes, may play a decisive role to control the inflammation.

The many different excited oxygen species produced by the activated phagocytes, for example, superoxide, peroxide, hydroxyl radical, alkoxyl radicals and singlet oxygen, are able to attack biologically important macromolecules, such as hyaluronic acid, proteins, DNA polynucleotides and lipid membranes[118–127]. The function of these biopolymers is expected to be considerably diminished. On the other hand, antigens may be formed which induce consecutive auto-immune reactions. Eventually, an aggravation of the inflammation may be observed (→ chronical inflammation)[97]. Peroxidation of membrane lipids proceeds in the same way, leading to severe damages of the cell[122–127]. Mutations can be induced by this process[121].

5.2 *Systemic Reaction*

When the damage has been sufficiently severe, the local inflammatory reaction is followed by metabolic changes affecting the entire organism[128, 129]:

- increase of leukocyte number (caused by stimulating the bone marrow),
- amino acid flux from the muscles to the liver (for energy formation and protein synthesis),
- drop of the albumin and transferrin levels in the serum,
- (increased) synthesis of the acute phase proteins in the liver (C-reactive protein, coeruloplasmin, α_1-antitrypsin, haptoglobin, fibrinogen) (for limitation of tissue damages, support of wound healing, modulation of immune functions[130]),
- rise of the serum copper level (up to 60 μM); the concentration of all copper fractions including coeruloplasmin, copper serum albumin and low molecular weight copper complexes is increased (see 3.1),
- a drop of the iron and zinc levels in the serum (absorption by the liver).

The systemic reaction aims to limit the damaging process by supporting the local reaction[130]. The supply of phagocytes is ascertained to facilitate resorption of damaged tissue and the scavenging of microorganisms; at the same time perished leukocytes are replaced. The amino acid metabolism is diverted to synthesize the acute phase proteins which are known to fulfil important functions in defence and repair mechanisms. The reduced serum iron concentration is hypothesized to inhibit the growth of iron-dependent microorganisms. In general, the levels of both zinc and copper in the serum are antagonistic. The anti-inflammatory activity of coeruloplasmin and copper complexes has already been discussed (see 4).

When a systemic reaction is triggered, the leukocyte endogenous mediator (LEM)[129] – which is produced by leukocytes – plays an essential function. In addition to LEM, other products of the arachidonic acid and/or superoxide metabolism are discussed to act as further mediators in the systemic reaction.

5.3 Inhibition of the Inflammation

There are several experimental approaches to examine the anti-inflammatory activity of a drug in animal model systems. The inflammatory process is induced using different means of irritation, for example, injections in the paw (paw edema) using carrageenan (an extract of algae), kaolin, dextran or urate[110, 131]. When a sterile cotton pad is implanted in a wound, the inflammation is accompanied by granuloma formation[132]. A model for the chronical inflammation is polyarthritis in rats. It is induced by injection of Freund's adjuvant, an emulsion containing denatured tuberculosis bacteria, paraffin and mannitol-monooleate[133]. The above tests were used to examine the many anti-inflammatorily active substances of biological and synthetical origin. Apart from the copper compounds (see 4), two pharmacologically important groups will be discussed in more detail: steroidal and non-steroidal anti-inflammatory drugs (NSAIDs). Both kinds of drugs affect the biosynthesis of prostaglandins. Unlike the glucocorticoids which inhibit the generation of arachidonic acid from membrane phospholipids (induction of the inhibitor of the phospholipase A_2)[134, 135], many non-steroidal drugs, including indomethacin and aspirin, inhibit the cyclooxygenase[136] which is known to convert arachidonic acid via the endoperoxide prostaglandin G (PGG) into the endoperoxide prostaglandin H (PGH). Other NSAIDs (for example benzoxaprofen[137]) inhibit the lipoxygenase pathway where leukotrienes are formed from arachidonic acid. Prostaglandins and leukotrienes are mediators of inflammation. The inhibition of their biosynthesis in the presence of steroidal and non-steroidal anti-inflammatory drugs explains their influence on the inflammatory process. Indomethacin, for example, inhibits vasodilatation, increase of the permeability of the blood vessels as well as leukocyte infiltration following an inflammatory stimulus[138–141]. In addition to the known inhibition of the prostaglandin biosynthesis, other mechanisms cannot be excluded. The possible participation of copper in the inhibition of cyclooxygenase awaits further examinations (see 6.1.3).

6 Mechanism of the Anti-Inflammatory Activity of Copper

6.1 Effects on the Initiation Phase of Inflammations

Following the tissue damage by an inflammatory impact (injury, microbial infection, immune reaction), the organism attempts to minimize those damages or bring them to a halt by infiltrating phagocytes, which are able to catabolize the foreign material (bacteria, destroyed cells). However, the secretion of lysosomal enzymes and reactive oxygen species may cause a secondary damage.

At the beginning of an inflammatory reaction, the so-called initiation phase, the anti-inflammatory activity of copper might be assigned to several possible actions: removal of superoxide by catalytically active low molecular weight copper chelates, the possible induction of anti-inflammatory copper proteins, influence of the prostaglandin biosynthesis or the immune reaction and microbicidal activity of copper.

6.1.1 Superoxide Dismutase Activity of Copper Complexes

The catalytical superoxide dismutase activity of differently chelated copper(I)/copper(II) complexes has been successfully shown. Formate[142)], amino acids[47, 143–146)], salicylates or derivatives thereof[147–149)], indomethacin and lonazolac and other derivatives of acetate[122, 150)], and penicillamine[151, 152)] are listed among the many potent copper-chelating ligands (Table 4). In the first step, copper(II) is reduced by superoxide to yield copper(I) which is reoxidized by another superoxide anion; thus the catalytical cycle is closed.

$$Cu(II) + \bullet O_2^- \rightarrow Cu(I) + O_2$$

$$Cu(I) + \bullet O_2^- \rightarrow Cu(II) + H_2O_2$$

Table 4. Superoxide dismutase activity of some copper complexes[122, 145, 146, 148)]. Superoxide dismutase activity was determined by pulse radiolysis employing a Febetron unit. A xenon lamp, Osram XBO 450 W 4, served as light source. In some of the experiments the optical detection system was composed of a Zeiss M 4 QIII monochromator, an EMI 9659 QB photomultiplier and a Textronix 7704 oscilloscope. Oscilloscope traces showing the decay of $\bullet O_2^-$ at 250 nm were photographed with a Polaroid camera. In another set of experiments the oscilloscope and the camera were replaced by transient recorder DATALAB 905. 40 ns pulses of high-energy electrons (1.81 MeV) were employed in oxygen saturated solutions of triply distilled and pyrolyzed water containing the copper complexes and 10^{-2} M formate at pH 7.5. The G-value of $\bullet O_2^-$ was raised thus to 6.1. Superoxide concentrations were estimated employing thiocyanate dosimetry[152)]. All rate constants represent the constants of second order reactions (k_2). In the cases of spontaneous superoxide dismutation, evaluation of the decay kinetics of superoxide radicals resulted in second order type reaction directly. When copper-penicillamine was present, the reactions showed pseudo-first order type kinetics, from which, depending on the concentration of the complex, the second order rate constants were calculated using a Wang 2200 computer. For further experimental details see Refs. 144, 145 and 148

Cu-complex	$k_2 \times 10^{-9}\ M^{-1}\ s^{-1}$
Cu_2Zn_2superoxide dismutase	1.30
Cu $(Lys)_2$	0.56
Cu $(Tyr)_2$	1.00
Cu $(Gly\text{-}His)_2$	0.29
Cu $(Gly\text{-}His\text{-}Leu)_2$	0.21
Cu $(formate)_2$	2.70
Cu $(salicylate)_2$	1.64
Cu $(acetylsalicylate)_2$	0.96
Cu $(diisopropylsalicylate)_2$	2.40
$Cu_2(indomethacin)_4$	3.00
Cu-penicillamine	0.40

As an example for a possible structure-activity relationship, the copper-indomethacin complex is presented in greater detail. X-ray diffraction studies revealed a unique structure of the copper-indomethacin complex (Figs. 8 and 9). Two centrosymmetrical dimers, $[Cu(indomethacin)_2]_2 \cdot L_2$, are seen. The dimerisation occurs in a bridging manner via four indomethacinate ions. Two coordination sites in the copper-copper axis are occupied by the solvent which can be readily replaced by superoxide. In the case of the $Cu_2(indomethacin)_2(Me_2SO)_2$, a third molecule is loosely bound in the unit cell. Either of the 3 d^9 coppers is located in a Jahn-Teller distorted octahedral surroundings. Four-square planar carboxyl oxygen atoms are 1.96 Å apart. The solvent oxygen (in the case of Me_2SO) is separated by 2.16 Å and the distance to the second copper atom is 2.63 Å. The copper-copper distance of 2.63 Å comes very close to that reported for metallic copper (2.56 Å). This supports very nicely the results obtained from the magnetic measurements where it was deduced that copper was antiferromagnetically coupled. It should be pointed out that the carboxylate bridges force a close contact of the two copper atoms. This is deduced from the narrow distance of only 2.23 Å (average) by looking at O(1)–O(2) and O(3)–O(4) (Fig. 9). Unlike the copper-copper distance, these binding lengths are much closer. Thus, each copper sticks 0.2 Å out of the square plane. The distance between the copper and the carboxylate oxygen are in agreement with the sum of the covalent radii of Cu and O (2.02 Å, see Pauling[153]).

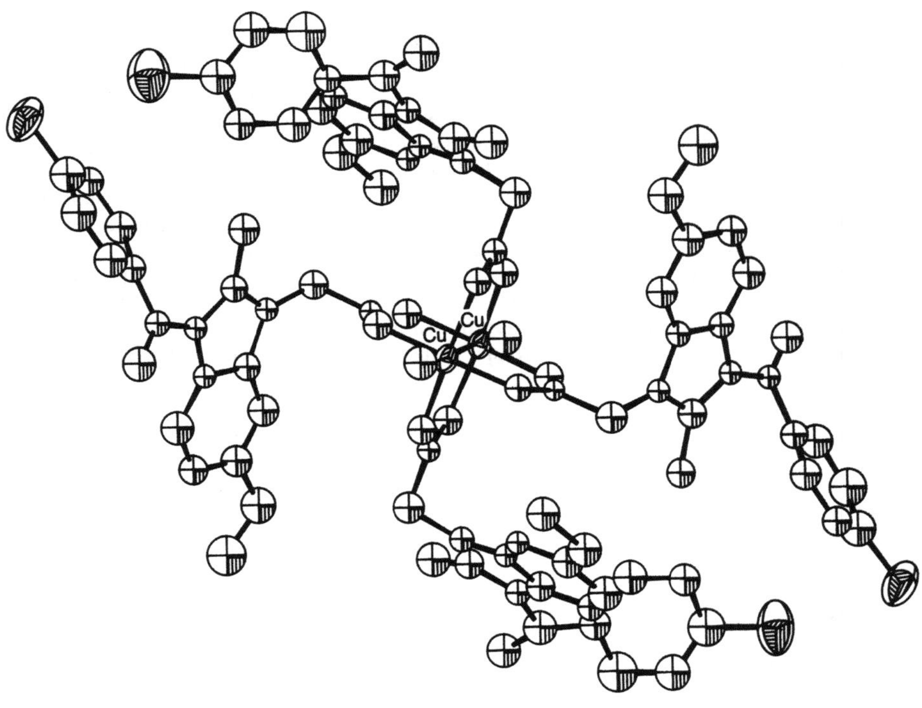

Fig. 8. Crystal structure of $Cu_2(indomethacin)_4 \cdot L_2$. (With permission from Biochimica et Biophysica Acta *631*, 232–245 (1980))

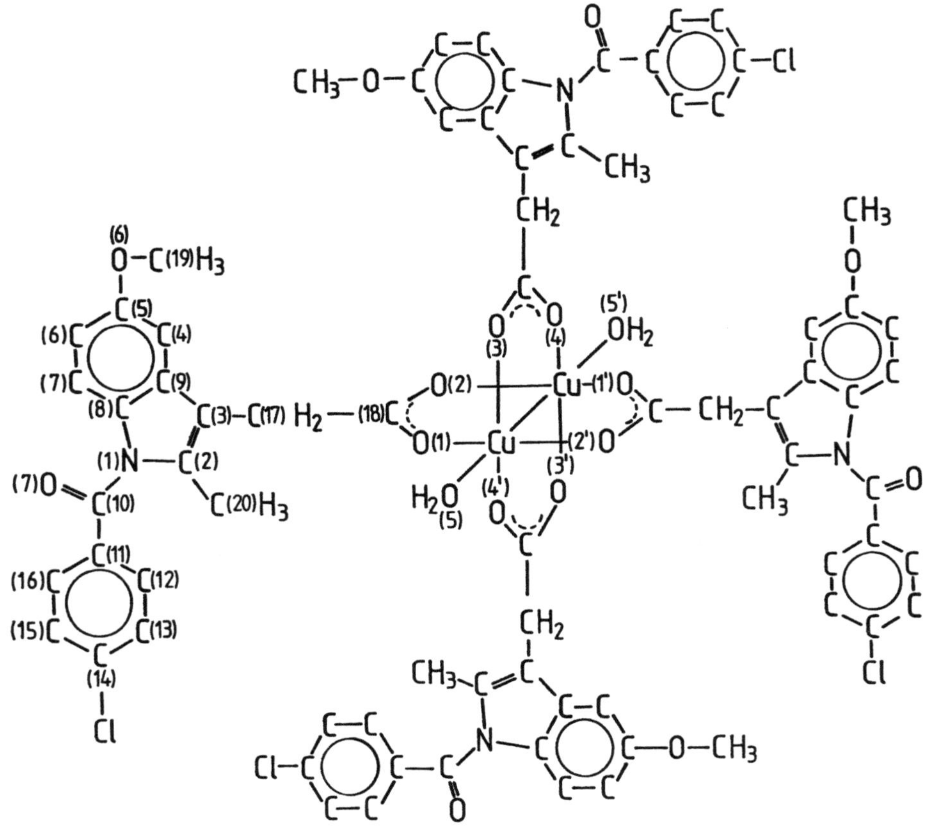

Fig. 9. Formula of Cu_2(indomethacin)$_4 \cdot L_2$. (With permission from Biochimica et Biophysica Acta *631*, 232–245 (1980))

The molecular architecture of Cu_2(indomethacin)$_4$ reaches back to a fairly widely distributed coordination of copper with oxygen, known as the dimeric copper acetate. Cu_2(acetate)$_4 \cdot H_2O$[154] is also antiferromagnetic, with a Néel point at − 18 °C[155]. This dimeric form of Cu(II)acetate is also present in Cu_2(indomethacin)$_4$. Furthermore, a dimeric Cu_2(salicylate)$_4$ of the same structural properties has been reported. The latter structure was deduced from electron paramagnetic measurements[156]. In addition, the same structure is known for Cu(II)formate · urea[157]. Throughout, two coordination sites in the copper-copper axis are occupied by some exchangeable ligand. Similarly, lonazolac, another substituted acetate derivative, was shown to form an antiferromagnetically coupled copper complex of possibly the same binuclear structure[150].

Electron transport between the two copper atoms should therefore be similar to that known for metallic copper. As pointed out earlier, this might explain the extraordinarily fast reaction with superoxide observed in all these copper complexes of acetate-like ligands. In the native Cu_2Zn_2superoxide dismutase, the copper-binding centres are separated by 34 Å and no direct electron transport is possible. In this case, electron transfer from one copper to the second, if any, must take place via the protein backbone, involving the histidine residues.

A predominant excited oxygen species generated by activated phagocytes is superoxide. This oxygen species is considered to be a mediator and trigger of many pathogenic events. Thus, it is very attractive to assign the anti-inflammatory activity of copper to its superoxide dismutase-mimicking activity. In the first place, the mediator effect of superoxide will be reduced. Formation of vasoactive and chemotactic factors and stimulation of the arachidonic acid metabolism will be decreased. On the other hand, a substantial reduction of damages induced by excited oxygen species derived from superoxide, including lipid peroxidation, catabolism of hyaluronic acid, unspecific reactions with proteins, formation of auto-antigens and mutagenicity (see 5.1), should be seen. The inflammatory process would proceed more mildly. An aggravation would be avoided.

Very little is known on the effect of copper compounds on these superoxide-dependent reactions in the organism. Some in vivo experiments support the superoxide-scavenging theory by copper complexes. For example, administered copper tyrosine inhibits the paracetamol-induced lipidperoxidation[158]. It is presumed that the superoxide dismutase activity of copper tyrosine inhibits the microsomal hydroxylation which is known to proceed via this oxygen radical. Consequently paracetamol is metabolized at a slower rate. The lipid peroxidation is attributed to the activity of free radicals which are generated during the metabolism of paracetamol.

6.1.2 Survival of Superoxide Dismutase Mimics in Biological Systems

It should be emphasized that the concentration of most of the low molecular weight copper complexes is extremely low in biological systems[150]. A detailed study on the reactivity of Cu_2(lonazolac)$_4$, a lipophilic copper acetate derivative, was performed to examine both the superoxide dismutase activity and survival in the presence of whole serum or serum albumin. Superoxide dismutase activity of the Cu-complex in micromolar concentrations was detectable in the presence of up to 900 μM per ml of serum albumin or whole serum protein. At 700 μM albumin concentration, a ternary complex between Cu_2(lonazolac)$_4$ and the protein was formed. The original acetate-copper coordination changed to a biuret-type copper bonding as seen from EPR and electron absorption spectrometry. Lonazolac did not induce a detectable conformational change of the protein near or at the copper binding site. Equilibrium dialysis and optical titration experiments revealed that essentially all copper of the Cu_2(lonazolac)$_4$ complex was bound in the specific binding site of serum albumin. The copper complex proved to be an effective inhibitor of lipid peroxidation.

According to the present results obtained from the reaction of serum albumin with copper chelates, not very much can be said about the actual concentration of superoxide dismutase-mimetic complexes in indirect assays for superoxide dismutase activity (for instance, inhibition of superoxide-driven nitro blue tetrazolium reduction). They contain many different metal chelators. For example, copper binding to xanthine oxidase protein (an enzyme used to generate superoxide in vitro) – similar to that observed for serum albumin – is possible. Numerous other chelators including nitro blue tetrazolium (NBT), gelatine and xanthine may react with the added copper complex. Thus, the actual concentration of the superoxide dismutase-mimicking complex is probably much lower.

It may come close to that of the 10 nM Cu_2Zn_2superoxide dismutase necessary to achieve 50% inhibition of NBT reduction. The assignment of the superoxide dismutase activity to "free Cu^{2+}" ions can be discarded. It has been shown that the free divalent copper ion concentration in complexes with amino acids or peptides lies between $10^{-12} - 10^{-18}$ M[47, 159].

At present, the following facts can be summarized. The acetate-like copper complexes are perfect mimics of superoxide dismutase activity in the absence of high concentrations of competitive chelating agents. In the presence of these ligands the concentration of intact complexes is very low. However, the absence of any superoxide dismutase activity at all cannot be ascertained. Some residual activity exerted by traces of undissociated copper complexes beyound the detection limit has to be considered. These traces may be sufficient to explain the anti-inflammatory action of many copper complexes observed in vivo[76].

Taking into account the 700 fold excess of serum albumin over the kinetically labile copper in serum, it is rather unlikely to expect an increase of superoxide dismutase activity following the external application of copper; essentially, all the copper is complexed by albumin in a form that has no superoxide dismutase activity. The extremely low superoxide dismutase activity in the extracellular compartments of the organism should be emphasized in this context. Respective values are: serum 0.6 U/ml, lymph 1.7 U/ml and are contrasted by the enormous activities in liver (1570 U/g wet weight) and erythrocytes (500 U/g wet weight)[160].

The anti-inflammatory response of known anti-inflammatory drugs, including salicylates, indomethacin, and lonazolac, may be assigned to the action of their copper-acetate-like clusters rather than the action of the free ligands[76, 161, 162]. Indeed, the catalysis of the spontaneous superoxide dismutation has been shown in water. In addition, the catalysis of these copper chelates dissolved in aqueous Me_2SO or acetonitrile proceeded in the same manner as in homogeneous aqueous systems. In contrast, the diminished reactivity of Cu_2Zn_2superoxide dismutase in aprotic solvents could probably be attributed to conformational changes of the protein moiety. For example, it is known that Me_2SO affects the protein portion of ferredoxin and Cu_2Zn_2superoxide dismutase[163, 164].

The power of the acetate-like superoxide dismutase mimics might be attributed to the lipophilicity of these complexes. Some copper complexes (especially those with lipophilic acetate derivatives) are soluble in lipid systems. Both the aromatic residues and the charge compensation following the reaction with divalent copper facilitate transport of the complex into and through lipophilic systems (Fig. 10). They could act as superoxide scavengers in lipophilic regions where Cu_2Zn_2superoxide dismutase is excluded.

6.1.3 Superoxide Dismutase-Active Complexes of High Stability

To improve possible applications of low molecular weight copper chelates provided with superoxide dismutase activity for analytical and/or therapeutical purposes, we were searching for a copper complex of higher resistance to strong chelators. With

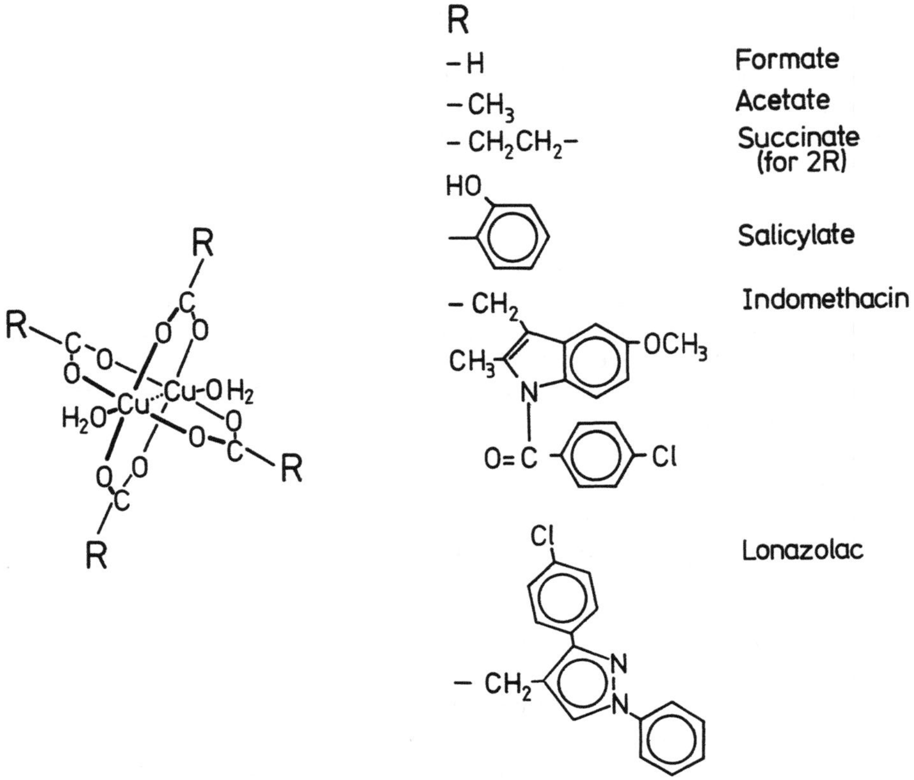

Fig. 10. Antiferromagnetically coupled and EPR-silent acetate-like Cu(II) complexes with increasing lipophilicity of the ligands. The antiferromagnetic coupling is also maintained in buffer-free aqueous solution. With increasing lipophilicity of the carboxylate ligand, the thermodynamic stability of the complex rises (see Table 3)

the exception of Cu-penicillamine and the native binuclear Cu_2Zn_2superoxide dismutase, all acetate-like copper complexes with NSAIDs did not survive EDTA treatment. EDTA was used to mimic the albumin fraction in the serum. Thus, the only low molecular weight copper complex having superoxide dismutase activity and known to be stable enough to survive treatment with serum albumin is the copper penicillamine complex. Penicillamine is generally used as a copper chelator in the case of copper poisoning and metabolic disorders of copper metabolism (Morbus Wilson). The copper is chelated and the complex is renally excreted.

In a pulse radiolysis study on the superoxide dismutase activity of Cu-penicillamine, the pH dependence and resistance to EDTA was examined in more detail. The second order rate constant of the spontaneous decay of $\bullet O_2^-$ was $2.3 \pm 0.3 \cdot 10^6\ M^{-1}s^{-1}$. In the presence of variable concentrations of Cu-penicillamine, markedly higher second order rate constants were determined. k_2 was highest ($6 \cdot 10^9\ M^{-1}s^{-1}$) using 10^{-8} M Cu-penicillamine and dropped to a constant value of $4.5 \pm 0.5 \cdot 10^8\ M^{-1}s^{-1}$ in the presence of micromolar or higher concentrations (Fig. 11).

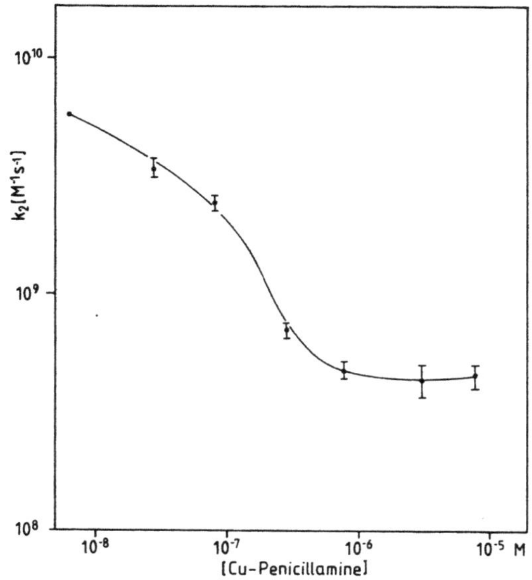

Fig. 11. The second order rate constant (k_2) of the reaction between Cu-penicillamine and $\cdot O_2^-$ at different concentrations of the copper chelate. Experimental details as in legend to Table 4. (With permission from Biochimica et Biophysica Acta *567*, 492–502 (1979))

It should be pointed out that the Cu-penicillamine-catalyzed superoxide dismutation proceeds with essentially the same velocity as observed with many known superoxide dismutases ($k_2 = 2.4 \cdot 10^9$ to $5.5 \cdot 10^8\ M^{-1}s^{-1}$)[165–170]. The significant rise of k_2 at $10^{-7} - 10^{-8}$ M concentrations of Cu-penicillamine could be attributed to the dissociation of extraneous Cu(II) from the outer sphere of the cluster. The disturbing reactivity should be countered by the addition of EDTA. This strong chelator is usually added to the different native superoxide dismutases to avoid uncontrolled side reactions of unspecifically bound transition metals (Fig. 12).

The chelator alone affected the $\cdot O_2^-$ decay. From 10^{-7} M on to higher concentrations of ethylene diamine tetraacetic acid suppressed the spontaneous $\cdot O_2^-$ dismutation. However, in the presence of $10^{-8} - 10^{-6}$ M Cu-penicillamine the catalysis of superoxide dismutation remained constant at $2 \pm 0.2 \cdot 10^8\ M^{-1}s^{-1}$. The apparent discrepancy between the data of Figs. 11 and 12 is due to an overlapping of the reaction of superoxide with the complex and the direct reaction of EDTA with $\cdot O_2^-$. This favours the assumption that dissociated spurious Cu(II) from the complex has been trapped by the chelator. The diminished rate constant caused by 10^{-6} M or higher EDTA concentrations must be assigned to the reactivity of free chelator and is not connected to the chelation of metal impurities.

The pH dependence of the Cu-penicillamine-catalyzed superoxide dismutation reveals no dramatic changes of k_2. Unlike the strong pH-dependent spontaneous reactions form $3 \cdot 10^7\ M^{-1}s^{-1}$ (pH 3) to $5 \cdot 10^5\ M^{-1}s^{-1}$ (pH 9), the same pH variations of Cu-penicillamine in the presence of EDTA solutions resulted in rate constants still being in the range of the usual enzymic catalyzed superoxide dismutation (Fig. 13).

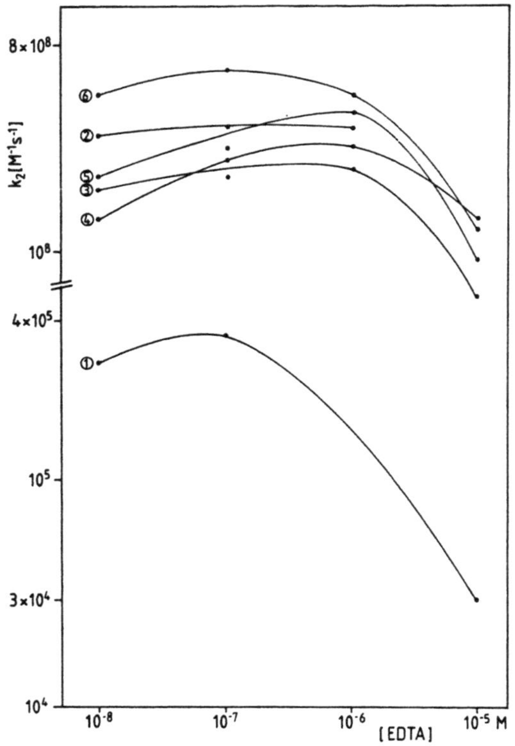

Fig. 12. Effect of different concentrations of EDTA on the rate constant of the reaction between Cu-penicillamine and $\bullet O_2^-$. 1, EDTA alone; 2–6 + Cu-penicillamine in the concentrations: 2, $9.1 \cdot 10^{-8}$ M; 3, $4.8 \cdot 10^{-7}$ M; 4, $9.1 \cdot 10^{-7}$ M; 5, $4.8 \cdot 10^{-6}$ M and 6, $9.1 \cdot 10^{-6}$ M. Higher inhibitor concentrations were not examined due to both the uncontrolled reactivity of the ligand with $\bullet O_2^-$ and the increase in viscosity of the solutions leading to erroneous rate constants. (With permission from Biochimica et Biophysica Acta *567*, 492–502 (1979))

6.1.4 The Superoxide Dismutase Activity in Vivo

Although the copper penicillamine complex exerts superoxide dismutase activity even in the presence of strong chelators, its reactivity in the organism must be seen in a different way. Penicillamine – when administered as an anti-rheumatoid drug – exerts its therapeutic effect only after several weeks of treatment. Thus, the assignment of the superoxide dismutase activity to the anti-rheumatoid activity is not easy to understand.

An important property of the superoxide dismutase-mimicking copper complexes is the need of at least one vacant coordination site to allow the reaction with superoxide radical. Thus, no superoxide dismutase activity is seen when EDTA, diethylene triaminopentaacetate, triethylene tetramine or the tripeptide Gly-Gly-His are chelated to Cu(II), forming complexes that are at least as stable as copper serum albumin.

The naturally occurring binuclear Cu_2Zn_2superoxide dismutase is an extraordinarily strong copper complex which has superoxide dismutase activity (Fig. 2). Its

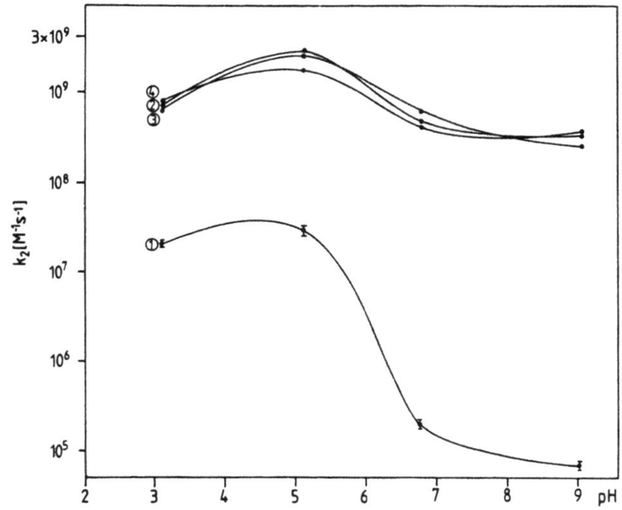

Fig. 13. pH-dependence of the rate constant of the reaction between superoxide and Cu-penicillamine on the presence of 10^{-6} M EDTA. Cu-penicillamine was present in the concentrations: 1, 0; 2, $4.8 \cdot 10^{-7}$ M; 3, $9.1 \cdot 10^{-7}$ M; 4, $4.8 \cdot 10^{-6}$ M. (With permission from Biochimica et Biophysica Acta *567,* 492–502 (1979))

physiological role is debated whether or not this enzyme controls superoxide-dependent mechanisms in the organism. Quite frequently it is used to demonstrate the occurrence of superoxide by inhibiting such a reaction in the presence of this copper-zinc-protein. Cu_2Zn_2superoxide dismutase is metabolized quite rapidly. The half life in blood is several minutes. Surprisingly, the reported anti-inflammatory activity is maintained much longer than the actual superoxide dismutase activity[94)]. Considerable doubts are raised against the hypothesis that the scavenging of superoxide is the most important mechanism of the anti-inflammatory reactivity of copper. Immune reactions of the protein portion or superoxide dismutase-active low molecular weight catabolic fragments of the native enzyme are discussed as the possible cause.

At present, we do not know whether or not the anti-inflammatory activity of copper is assigned to the superoxide dismutase activity or from copper chelates formed in vivo. The direct proof of superoxide dismutase activity under in vivo conditions is still awaited with great interest. Taking into account the extraordinarily low concentration of the low molecular weight copper chelates in aqueous biological systems, there is a fairly high probability that the superoxide dismutase activity of these complexes plays a minor role. However, in lipophilic sections of the organisms, including membranes and other lipophilic systems, the resulting copper complexes would have a better chance to survive and would be able to protect the membranes from the attack of excited oxygen species.

6.1.5 Induction of Cu_2Zn_2superoxide Dismutase

It would be attractive to conclude that an increased copper supplement parallels an increase of the biosynthesis of Cu_2Zn_2superoxide dismutase[119)] with the consequence of a diminished superoxide liberation from phagocytes. Rats and chickens which were maintained on a copper-deficient diet displayed a diminished superoxide dismutase activity in erythrocytes[171, 172]. By way of contrast, excessive copper administration caused only an insignificant rise in the enzymic activities (0–10%) compared to the animals maintained on a normal diet[171, 173]. These observations allow the conclusion that an induction of Cu_2Zn_2superoxide dismutase is only seen in copper-deficient animals following the repletion with copper.

During the copper repletion, the activity of Cu_2Zn_2superoxide dismutase rose within 36 h to 94% and remained at least 12 days at this level. Additional intraperitoneal copper doses to rats maintained on a full diet were unable to stimulate the Cu_2Zn_2superoxide dismutase activity in the blood significantly (11% above control) (Fig. 14).

6.1.6 Modulation of Prostaglandin Synthesis

Prostaglandins belong to the mediators of inflammation. The major portion is synthesized by activated macrophages and released in the environment. In an inflammatory event, the prostaglandins of the E-type (PGE) are of great importance. However, in the course of an inflammation they act contradictorily. Proinflammatory

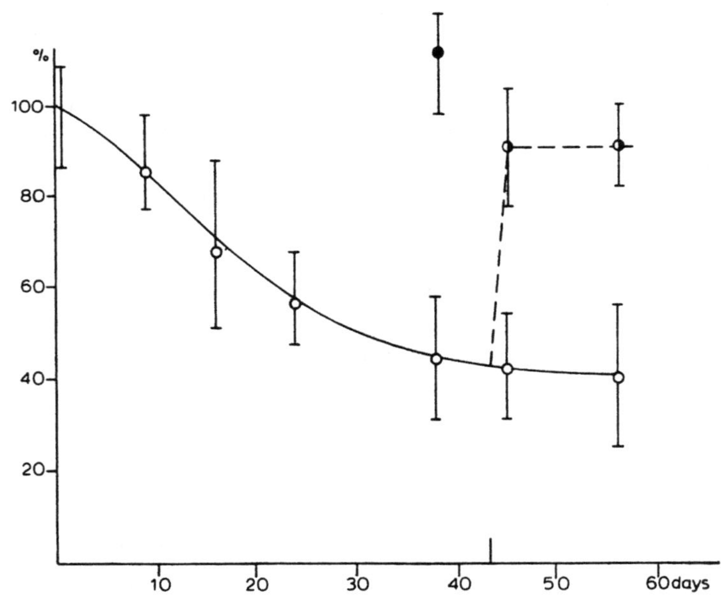

Fig. 14. Cu_2Zn_2superoxide dismutase activity in blood cells of copper-deficient rats in percent of the control. The single point at the 39th day represents the Cu_2Zn_2superoxide dismutase activity in the blood cells of rats (same age as the control), but after a 4-week treatment with seven intraperitoneal injections of 0.13 mg $CuSO_4$ in saline solution per kg weight. The significance on the 15th day between control and copper-deficient rats is better than $P < 0.001$. The dotted line indicates the Cu_2Zn_2superoxide dismutase activity in blood cells of rats having a dietary copper repletion via drinking water. (With permission from Biochimica et Biophysica Acta *444*, 396–406 (1976))

effects are vasodilatation, increased vasopermeability and aggravation of pain induced by other mediators like kinins, histamine. On the other hand, the suppression of both metabolism and function of leukocytes, especially lymphocytes, results in diminished inflammation (inhibition of chemotaxis and phagocytosis, abolishment of the activation of macrophages by lymphocytes)[174]. In other words, formation of edema and pain response are increased while the cellular defence mechanisms are inhibited. The discrepancy between either mode of action could be minimized in that pro- and anti-inflammatory effects occur in different target cells and at different times. The increased inflammatory response proceeds rather quickly and is especially seen in non-leukocyte cells (for example the cells of the vessel wall). The inhibitory action is seen at a later time and deals with leukocytes. There is a break of the positive feed back between macrophage and lymphocytes with the consequence of the regulation of the inflammatory process[175]. In contrast to the prostaglandins of the E series, $PGF_{2\alpha}$ is vasoconstrictive and, thus, anti-inflammatorily active. As for the rest, $PGF_{2\alpha}$ behaves alsmost neutral in relation to the inflammatory process.

In vitro, the prostaglandin biosynthesis is influenced by copper salts and other copper compounds[176–180]. On the whole, the formation of prostaglandins is reduced. At the same time, the production of $PGF_{2\alpha}$ is stimulated relative to PGE. It may be deduced that copper shifts the equilibrium between different prostaglandins to favour the anti-inflammatorily active $PGF_{2\alpha}$. The mechanism of this phenomenon is unknown.

6.1.7 Modulation of Immunobiological Functions

A biologically active immune system is an important condition for the successful defence against infectious microorganisms. Copper appears to be essential for the normal functioning of the immune system. In copper-deficient animals the number of antibody-forming cells is substantially reduced[181, 182]. Lymphocytes react more weakly on the stimulation of cell proliferation[183]. Lymphocytes treated with diethyldithiocarbamate synthesize less DNA compared to the control. This phenomenon can be reversed following the addition of inorganic copper salts[184]. The reaction of T-lymphocytes on mitogens (→ DNA-synthesis) is affected by D-penicillamine in the presence of copper salts[185]. The reactivity of copper penicillamine on lymphocytes is not known. At present, it may not be deduced whether or not copper is able to modulate the function of the immune cells in a way other than via its biochemical action in the many known copper proteins.

6.1.8 Antimycoplasmic Activity

Bacteria of the mycoplasmic group may be important in the etiology of rheumatic diseases[186]. Copper complexes of substituted phenantrolines and isoquinolines show a strong antimycoplasmic activity in vitro[187].

However, these complexes are very toxic. Perhaps there are endogenous copper complexes of similar reactivity. It should be noticed that copper sulphate and some copper complexes have a mycoplasmicidic effect in vitro[186].

6.2 Response in the Repair Phase

In an inflammatory event, the initiation phase overlaps with the repair phase, i.e. immediately after the occurrence of tissue damages the repair is started. In the case that the repair mechanisms are out of order, an aggravation of the inflammation might be the cause, as the initial damages are not removed. This results in a progressive increase of the total damages (secretion of lysosomal enzymes and reactive oxygen species).

Granulomas obtained following the insertion of cotton wad in an incision are of considerably superior quality than controls provided the animals have been treated with anti-inflammatory copper complexes. The new tissue was well formed and less edematous[76, 188]. Following the addition of copper an inflammation appears to be much milder. Not only edema formation and leukocyte infiltration are reduced, but also formation of granulation tissue is much more ordered. In this context, the biosynthesis of the extracellular substance of the connective tissues plays a decisive role.

There are two important processes of the repair phase which depend on copper: collagen maturation (lysyloxidase) and formation of new blood vessels in the inflamed region (angiogenesis).

6.2.1 Induction of Lysyloxidase

Fully matured collagen, a major portion of the extracellular component of the connective tissues, requires the enzyme lysyloxidase. This enzyme crosslinks different polypeptide chains of procollagen by an oxidative link between two lysyl side chains (ε-NH_2-groups):

$$R{-}CH_2{-}NH_2 + NH_2{-}CH_2{-}R + O_2 \rightarrow R{-}CH{=}N{-}CH_2{-}R + NH_3 + H_2O_2$$

A diminished activity of lysyloxidase, therefore parallels an insufficient repair of the tissue damage. The lysyloxidase activity was diminished in copper-deficient chicks[189]. It is not known whether copper supplementation to normally fed animals induces the activity of lysyloxidase.

6.2.2 Angiogenesis

Owing to the formation of new blood vessels (especially capillaries), there is a positive influence on the supply of the damaged region and its repair. Copper promises to actively induce angiogenesis. For example, copper-deficient rabbits are unable to respond to an angiogenetic reaction using adequate stimuli such as PGE_1 or neoplasmic cells[190]. Employing rabbit cornea preparations, added copper is capable of triggering angiogenesis. Low molecular weight copper chelates from heparin, Gly-His-Lys as well as coeruloplasmin or copper-containing fragments thereof, have been successfully used[71]. The accumulation of copper in the cornea following the PGE_1 induced angiogenesis should be mentioned. Copper-heparin is able to mobilize endothelial cells[191, 192]. A possible beneficial role of copper to maintain the capillary network in good shape is discussed.

7 Conclusion

The remarkable rise of mammalian serum copper is the most intriguing phenomenon in an inflammatory event induced either mechanically, microbially or immunologically. Externally applied copper supports this rise of serum copper. The copper-storage proteins of the organism do not need to release copper. Provided the rise in serum copper can be considered an endogenous anti-inflammatory reaction for limitation of potentially damaging mechanisms, the anti-inflammatory activity of many copper compounds may be seen in the support of the above-mentioned endogenous response. This conclusion is supported by an observed aggravation of inflammation in copper-deficient rats; they are unable to raise their serum copper levels significantly[97, 193].

Copper administration might also overcome a latent nutritional copper deficiency caused by insufficient nutritional copper supply. While in the healthy organism usually no symptoms are manifest, inflammatory and/or other diseases where copper is mobilized may be aggravated when copper stores are almost empty. This is even more pronounced when the excretion of copper and other trace elements is elevated in chronic inflammatory diseases. Investigations carried out in the United States showed that a lot of people consume less than the recommended 2 mg Cu/day[194] necessary to replace excreted copper. Reports from other industrial countries led to similar results.

The nutritional copper intake deserves more attention. The possible relationship between dietary copper and the severity of the inflammatory process, especially in the light of copper deficiency, is unknown. Nevertheless, copper supplementation can be recommended for treatment of inflammatory diseases as it will facilitate to maintain the optimal serum copper level. The use of copper complexes should not be over-estimated to be of dramatic value, although there might be a beneficial therapeutic effect.

The elucidation of the mode of action of the anti-inflammatory activity of copper complexes awaits further studies. The earlier exclusive assignment of the anti-inflammatory activity of these complexes to their superoxide dismutase-like activity has to stand up to rigorous questioning. At present, no direct proof of the scavenging of excited oxygen species in vivo is available. Only in vitro studies on the effect of copper on the biosynthesis of prostaglandins and the metabolism of other hormone mediators have been performed. There may be other metabolic pathways which are dependent on copper. The recently discovered beneficial effect of copper on angiogenesis is a promising start in this direction.

Acknowledgements

Portions of the experimental part were supported by grants from the Deutsche Forschungsgemeinschaft (We 401/17) and the Fonds der Chemischen Industrie. Thanks for technical help in preparing the manuscript go to D. Schell.

8 References

1. Joachim, H.: Papyros Ebers, Berlin, Georg Reimer Verlag 1890
2. Plinius der Ältere, C. P.: Naturalis Historia, Liber XXXIV, Stuttgart, Teubner Verlag 1967
3. Rademacher, J. G.: Rechtfertigung der von den Gelehrten mißkannten, verstandesrechten Erfahrungsheillehre der alten scheidekünstigen Geheimärzte und treue Mittheilung der Ergebnisse einer 25jährigen Erprobung dieser Lehre am Krankenbette, Berlin, Georg Reimer Verlag 1843
4. Kobert, R.: Dtsch. Med. Wochenstr. *21,* 5 (1895)
5. Hart, E. B., Steinbock, H., Waddell, J., Elvehjem, C. A.: J. Biol. Chem. *77,* 797 (1928)
6. Basolo, F., Pearson, R. G.: Mechanisms of Inorganic Reactions, 2nd edition, New York, J. Wiley 1967
7. Lontie, R., Witters, R.: Inorganic Biochemistry (ed. Eichhorn, G.) p. 344, Amsterdam, Elsevier 1973
8. Weser, U.: Struct. Bonding *17,* 1 (1973)
9. Fridovich, I.: Annu. Rev. Biochem. *44,* 147 (1975)
10. Fee, J. A.: Metal Ions in Biological Systems, Vol. 13 (ed. Sigel, H.) p. 259, New York, Marcel Dekker 1981
11. Büttner, G. R.: Inflammatory Diseases and Copper (ed. Sorenson, J. R. J.) p. 199, Clifton NJ, Humana Press 1982
12. Harris, E. D., DiSilvestro, R. A., Balthrop, J. E.: ibid., p. 183
13. Opsahl, W.: J. Nutr. *112,* 708 (1982)
14. Knowles, P. F., Lowe, D. J., Peters, J., Thorneley, R. N. F., Yadav, K. D. S.: NATO Adv. Study Inst. Ser., Ser. C *100,* 159 (1983)
15. Dooley, D. M., Golnik, K. C.: J. Biol. Chem. *258,* 4245 (1983)
16. Eipper, B. A., Mains, R. E., Glembotski, C. C.: Proc. Natl. Acad. Sci. U.S.A. *80,* 5144 (1983)
17. Lerch, K.: Metal Ions in Biological Systems, Vol. 13 (ed. Sigel, H.) p. 143, New York, Marcel Dekker 1981
18. Skotland, T., Flatmark, T.: Eur. J. Biochem. *132,* 171 (1983)
19. Brunori, M., Antonini, E., Wilson, M. T.: Metal Ions in Biological Systems, Vol. 13 (ed. Sigel, H.) p. 187, New York, Marcel Dekker 1981
20. Kadenbach, B.: Angew. Chem. *95,* 273 (1983)
21. Fee, J. A.: Struct. Bonding *23,* 1 (1975)
22. Frieden, E.: Metal Ions in Biological Systems, Vol. 13 (ed. Sigel, H.) p. 117, New York, Marcel Dekker 1981
23. Goldstein, I. M., Charo, I. F.: Lymphokines *8,* 373 (1983)
24. Lappin, A. G.: Metal Ions in Biological Systems, Vol. 13 (ed. Sigel, H.) p. 15, New York, Marcel Dekker 1981
25. Hartmann, H. J., Weser, U.: Biochim. Biophys. Acta *491,* 211 (1977)
26. Brady, F. O.: Trends Biochem. Sci. *7,* 143 (1982)
27. Weser, U., Hartmann, H. J.: Copper Proteins and Copper Enzymes, Vol. III (ed. Lontie, R.) p. 151, Boca Raton, CRC Press 1984
28. Tainer, J. A., Getzoff, E. D., Richardson, J. S., Richardson, D. C.: 250 D: Cu, Zn-Superoxide Dismutase Complete Atomic Coordinates (eds. Richardson, D. C., Richardson, J. S.) Brookhaven NY, Brookhaven Protein Structure Data Bank 1980
29. Malkin, R., Malmström, B. G.: Adv. Enzymol. *33,* 177 (1970)
30. Gray, H. B.: Adv. Inorg. Biochem. *2,* 1 (1980)
31. Gray, H. B., Solomon, E. I.: Copper Proteins, Metal Ions in Biology, Vol. 3 (ed. Spiro, T. G.) p. 1, New York, John Wiley 1981
32. Solomon, E. I., Hare, J. W., Dooley, D. M., Dawson, J. H., Stephens, P. J., Gray, H. B.: J. Am. Chem. Soc. *102,* 168 (1980)
33. Thompson, J. S., Marks, T. J., Ibers, J. A.: ibid. *101,* 4180 (1979)
34. Co, M. S., Hodgson, K. O., Eccles, T. K., Lontie, R.: ibid. *103,* 984 (1981)

35. Hahn, J. E., Co, M. S., Spira, D. J., Hodgson, K. O., Solomon, E. I.: Biochem. Biophys. Res. Commun. *112,* 737 (1983)
36. Spira, D. J., Co, M. S., Solomon, E. I., Hodgson, K. O.: ibid. *112,* 746 (1983)
37. Gubler, C. J., Lahey, M. E., Cartwright, G. E., Wintrobe, M.: J. Clin. Invest. *32,* 405 (1953)
38. Sternlieb, I., Morell, A. G., Tucker, W. D., Greene, M. W., Scheinberg, I. H.: ibid. *40,* 1834 (1961)
39. Evans, G. W.: Physiol. Rev. *53,* 535 (1973)
40. Lau, S., Sarkar, B.: J. Biol. Chem. *246,* 5938 (1971)
41. Earl, C. J., Moulton, M. J., Selverstone, B.: Am. J. Med. *17,* 205 (1954)
42. Neumann, P. Z., Sass-Kortsak, A.: J. Clin. Invest. *46,* 646 (1967)
43. May, P. M., Williams, D. R.: Metal Ions in Biological Systems, Vol. 12 (ed. Sigel, H.) p. 283, New York, Marcel Dekker 1981
44. Klug-Roth, D., Rabani, J.: J. Phys. Chem. *80,* 588 (1976)
45. Yatsimirski, K. B., Vasilev, V. P.: Instability Constants of Complex Compounds, p. 149, Oxford, Pergamon Press 1960
46. Abbasi, S. A., Sharma, R. K.: J. Inorg. Nucl. Chem. *43,* 625 (1981)
47. Weinstein, J., Bielski, B. H. J.: J. Am. Chem. Soc. *102,* 4916 (1980)
48. Sarkar, B.: Metal Ions in Biological Systems, Vol. 12 (ed. Sigel, H.) p. 233, New York, Marcel Dekker 1981
49. Jackson, G. E., May, P. M., Williams, D. R.: FEBS Lett. *90,* 173 (1978)
50. Bearn, A. G., Kunkel, H. G.: Proc. Soc. Exp. Biol. *85,* 44 (1954)
51. Stein, W. H., Moore, S.: J. Biol. Chem. *211,* 915 (1954)
52. Felig, P.: Annu. Rev. Biochem. *44,* 933 (1975)
53. Sorenson, J. R. J.: Prog. Med. Chem. *15,* 211 (1978)
54. Niedermeier, W.: Ann. Rheum. Dis. *24,* 544 (1965)
55. Lorber, A., Cutler, L. S., Chang, C. C.: Arthritis Rheum. *11,* 65 (1968)
56. Kampschmidt, R. F.: J. Reticuloendothelial Soc. *23,* 287 (1978)
57. Koskelo, P., Kekki, M., Nikkila, E. A., Virkkunen, M.: Scand. J. Clin. Lab. Invest. *19,* 259 (1967)
58. Hartmann, H. J., Morpurgo, L., Desideri, A., Rotilio, G., Weser, U.: FEBS Lett. *152,* 94 (1983)
59. Morpurgo, L., Hartmann, H. J., Desideri, A., Weser, U., Rotilio, G.: Biochem. J. *211,* 515 (1983)
60. Hartmann, H. J., Gärtner, A., Weser, U.: Hoppe-Seyler's Z. Physiol. Chem. *365,* 1355 (1984)
61. Sass-Kortsak, A.: Adv. Clin. Chem. *8,* 1 (1965)
62. Neumann, P. Z., Silverberg, M.: Nature *210,* 414 (1966)
63. Harris, D. I. M., Sass-Kortsak, A.: J. Clin. Invest. *46,* 659 (1967)
64. Neumann, P. Z., Silverberg, M.: Nature *213,* 775 (1967)
65. Pickart, L., Freedman, J. H., Loker, W. J., Peisach, J., Perkins, C. M., Stenkamp, R. E., Weinstein, B.: ibid. *288,* 715 (1980)
66. Schmitt, R. C., Darwish, H. M., Cheney, J. C., Ettinger, M. J.: Am. J. Physiol. *244,* G 183 (1983)
67. Smith, B. S. W., Wright, H.: Biochim. Biophys. Acta *307,* 590 (1973)
68. Marceau, N., Aspin, N.: Am. J. Physiol. *222,* 106 (1972)
69. Hsieh, H. S., Frieden, E.: Biochem. Biophys. Res. Commun. *67,* 1326 (1975)
70. Harris, E. D., DiSilvestro, R. A.: Proc. Soc. Exp. Biol. Med. *166,* 528 (1981)
71. Raju, K. S., Alessandri, G., Ziche, M., Gullino, P. M.: J. Natl. Cancer Inst. *69,* 1183 (1982)
72. Owen, C. A.: Proc. Soc. Exp. Biol. Med. *149,* 681 (1975)
73. Fenz, E.: Muench. Med. Wochenschr. *41,* 1101 (1951)
74. Forestier, J., Certonciny, A.: Presse Med. *64,* 884 (1946)
75. Hangarter, W. S.: Inflammatory Diseases and Copper (ed. Sorenson, J. R. J.) p. 439, Clifton NJ, Humana Press 1982
76. Sorenson, J. R. J.: J. Med. Chem. *19,* 135 (1976)
77. Rainsford, K. D., Whitehouse, M. W.: J. Pharm. Pharmacol. *28,* 83 (1976)
78. Denko, C. W., Whitehouse, M. W.: J. Rheumatol. *3,* 54 (1976)
79. Whitehouse, M. W., Walker, W. R.: Agents Actions *8,* 85 (1978)
80. Lewis, A. J.: ibid. *8,* 244 (1978)

81. Williams, D. A., Walz, D. T., Foye, W. O.: J. Pharm. Sci. *65,* 126 (1976)
82. Lewis, A. J., Smith, W. E., Brown, D. H.: Agents Actions Suppl. *8,* 327 (1981)
83. Lewis, E. J., West, G. B.: ibid. *8,* 339 (1981)
84. Perrin, D. D., Whitehouse, M. W.: ibid. *8,* 261 (1981)
85. Walker, W. R., Beveridge, S. J., Whitehouse, M. W.: Agents Actions *10,* 38 (1980)
86. Beveridge, S. J., Walker, W. R., Whitehouse, M. W.: J. Pharm. Pharmacol. *32,* 425 (1980)
87. Dollwet, H. H. A., Schmidt, S. P., Seeman, R. E.: Inflammatory Diseases and Copper (ed. Sorenson, J. R. J.) p. 347, Clifton NJ, Humana Press 1982
88. Thiery, M.: Contracept. Delivery Syst. *4,* 175 (1983)
89. Walker, W. R., Griffin, B. J.: Search *7,* 100 (1976)
90. Walker, W. R., Reeves, R. R., Brosnan, M., Coleman, G. D.: Bioinorg. Chem. *7,* 271 (1977)
91. Oyanagui, Y.: Biochem. Pharmacol. *25,* 1465 (1976)
92. McCord, J. M., Stokes, S. M., Wong, K.: Adv. Inflammation Res. *1,* 273 (1979)
93. Wolf, B.: Inflammatory Diseases and Copper (ed. Sorenson, J. R. J.) p. 453, Clifton NJ, Humana Press 1982
94. Baret, A., Jadot, G., Valli, M., Bruguerolle, B., Puget, K., Michelson, A. M.: Oxy Radicals and Their Scavenger Systems, Volume II (ed. Cohen, G., Greenwald, R.A.) p. 274, New York NY, Elsevier 1983
95. Denko, C. W.: Agents Actions *9,* 333 (1979)
96. Milanino, R., Conforti, A., Fracasso, M. E., Franco, L., Leone, R., Passarella, A., Tarter, G., Velo, G. P.: ibid. *9,* 581 (1979)
97. Milanino, R., Velo, G. P.: Agents Actions Suppl. *8,* 209 (1981)
98. Vane, J. R.: Angew. Chem. *95,* 782 (1983)
99. Samuelsson, B.: Angew. Chem. *95,* 854 (1983)
100. Bergström, S.: Angew. Chem. *95,* 865 (1983)
101. Hammarström, S.: Annu. Rev. Biochem. *52,* 355 (1983)
102. Pinckard, R. N.: Monogr. Pathol. *23,* 38 (1982)
103. Salin, M. L., McCord, J. M.: Superoxide and Superoxide Dismutases (eds. Michelson, A. M., McCord, J. M., Fridovich, I.) p. 257, New York, Academic Press 1977
104. Badwey, J. A., Karnovsky, M. L.: Annu. Rev. Biochem. *49,* 695 (1980)
105. Markert, M., Allaz, M. J., Frei, J.: FEBS Lett. *113,* 225 (1980)
106. Kuehl, F. A., Humes, J. L., Egan, R. W., Ham, E. A., Beveridge, G. C., van Arman, C. G.: Nature *265,* 170 (1977)
107. Singh, A.: Photochem. Photobiol. *28,* 429 (1978)
108. Paschen, W., Weser, U.: Hoppe-Seyler Z. Physiol. Chem. *356,* 727 (1975)
109. Janoff, A., Harvey, C.: Monogr. Pathol. *23,* 62 (1982)
110. Vinegar, R., Schreiber, W., Hugo, R.: J. Pharmacol. Exp. Ther. *166,* 96 (1969)
111. Bonta, I. L., Bult, H., Parnham, M. J., Vincent, J. E.: Agent Actions *8,* 98 (1978)
112. Flower, R. J., Harvey, E. A., Kingston, W. P.: Br. J. Pharmacol. *56,* 229 (1976)
113. DelMaestro, R. F., Björk, A., Arfors, K. E.: Microvasc. Res. *22,* 239 (1981)
114. Bitter-Suermann, D.: Agents Actions Suppl. *11,* 159 (1982)
115. Del Principe, D., Menichelli, A., Galli, E., Persiani, M., Perlini, M., D'Arcangelo, C., Businco, L., Rossi, P.: Pediatr. Res. *16,* 1000 (1982)
116. McCord, J. M., Petrone, W. F.: Lipid Peroxides in Biology and Medicine, Proc. Int. Conf. 1980 (ed. Yagi, K.) p. 123, New York, Academic Press 1982
117. Halliwell, B.: Copper Proteins and Copper Enzymes Vol. II (ed. Lontie, R.) p. 63, Boca Raton CRC Press 1984
118. Brawn, K., Fridovich, I.: Acta Physiol. Scand. Suppl. *492,* 9 (1980)
119. McCord, J. M.: Science *185,* 529 (1974)
120. Lavelle, F., Michelson, A. M., Dimitrijevic, L.: Biochem. Biophys. Res. Commun. *55,* 350 (1973)
121. Moody, C. S., Hassan, H. M.: Proc. Natl. Acad. Sci. U.S.A. *79,* 2855 (1982)
122. Weser, U., Sellinger, K. H., Lengfelder, E., Werner, W., Strähle, J.: Biochim. Biophys. Acta *631,* 232 (1980)
123. Kellogg III, E. W., Fridovich, I.: J. Biol. Chem. *250,* 8812 (1975)
124. Thomas, M. J., Mehl, K. S., Pryor, W. A.: Biochem. Biophys. Res. Commun. *83,* 927 (1978)
125. Gutteridge, J. M. C.: ibid. *77,* 379 (1977)

126. Lynch, R. E., Fridovich, I.: J. Biol. Chem. *253,* 1838 (1978)
127. Goldberg, B., Stern, A.: Arch. Biochem. Biophys. *178,* 218 (1977)
128. Powanda, M. C.: Agents Actions Suppl. *8,* 121 (1981)
129. Powanda, M. C.: Inflammatory Diseases and Copper (ed. Sorenson, J. R. J.) p. 31, Clifton NJ, Humana Press 1982
130. Powanda, M. C.: Am. J. Clin. Nutr. *30,* 1254 (1977)
131. Winter, C. A., Risley, E. A., Nuss, G. W.: J. Pharmacol. Exp. Ther. *141,* 369 (1963)
132. Meier, R., Schuler, W., Desaulles, P.: Experientia *6,* 469 (1950)
133. Aspinall, R. L., Cammarata, P. S.: Nature *224,* 1320 (1969)
134. Flower, R. J., Blackwell, G. P.: ibid. *278,* 456 (1979)
135. Hirata, F., Schiffmann, E., Venkatasubramanian, K., Salomon, D., Axelrod, J.: Proc. Natl. Acad. Sci. U.S.A. *77,* 2533 (1980)
136. Shen, T. Y.: Handb. Exp. Pharmacol. *50,* 305 (1979)
137. Walker, J. R., Dawson, W.: J. Pharm. Pharmacol. *31,* 778 (1979)
138. Issekutz, A. C.: Immunopharmacol. *5,* 183 (1983)
139. Bailey, P. J.: Biochem. Pharmacol. *32,* 1213 (1982)
140. Yamashita, T.: Inflammation *6,* 87 (1982)
141. Miyasaka, K.: Eur. J. Pharmacol. *77,* 229 (1982)
142. Rabani, J., Klug-Roth, D., Lilie, J.: J. Phys. Chem. *77,* 1169 (1973)
143. Joester, K. E., Jung, G., Weber, U., Weser, U.: FEBS Lett. *25,* 25 (1972)
144. Brigelius, R., Spöttl, R., Bors, W., Lengfelder, E., Saran, M., Weser, U.: ibid. *47,* 72 (1974)
145. Brigelius, R., Hartmann, H. J., Bors, W., Saran, M., Lengfelder, E., Weser, U.: Hoppe-Seyler Z. Physiol. Chem. *356,* 739 (1975)
146. Weser, U., Schubotz, L. M.: Agents Actions Suppl. *8,* 103 (1981)
147. DeAlvare, L. R., Goda, K., Kimura, T.: Biochem. Biophys. Res. Commun. *69,* 687 (1976)
148. Younes, M., Lengfelder, E., Zienau, S., Weser, U.: ibid. *81,* 576 (1978)
149. O'Young, C. L., Lippard, S. J.: J. Am. Chem. Soc. *102,* 4920 (1980)
150. Deuschle, U., Weser, U.: Inorg. Chim. Acta *91,* 237 (1984)
151. Younes, M., Weser, U.: Biochem. Biophys. Res. Commun. *78,* 1247 (1977)
152. Lengfelder, E., Fuchs, C., Younes, M., Weser, U.: Biochim. Biophys. Acta *567,* 492 (1979)
153. Pauling, L.: The Nature of the Chemical Bond, Ithaca NY, 1960
154. Van Niekerk, J. N., Schoening, F. R. L.: Acta Crystallogr. *6,* 227 (1953)
155. Figgis, B. N., Martin, R. L.: J. Chem. Soc. 3837 (1956)
156. Suntsov, E. V., Ablov, A. V., Popovich, G. A., Kiosse, G. A.: Dokl. Akad. Nauk. S. S. R. *179,* 1352 (1968)
157. Yawney, D. B. W., Doedens, R. J.: Inorg. Chem. *9,* 1626 (1970)
158. Wendel, A., Heidinger, S.: Res. Commun. Chem. Pathol. Pharmacol. *28,* 473 (1980)
159. Amar, C., Vilkas, E., Foos, J.: J. Inorg. Biochem. *17,* 313 (1982)
160. Marklund, S.: Acta Physiol. Scand. Suppl. *492,* 19 (1980)
161. Oyanagui, Y.: Biochem. Pharmacol. *19,* 135 (1976)
162. Sorenson, J. R. J.: Inflammation *1,* 317 (1976)
163. Cammack, R.: Biochem. Biophys. Res. Commun. *54,* 548 (1973)
164. Symonyan, M. A., Nalbandyan, R. M.: Biochim. Biophys. Acta *446,* 432 (1976)
165. Rotilio, G., Bray, R. C., Fielden, E. M.: ibid. *268,* 605 (1972)
166. Klug, D., Rabani, J., Fridovich, I.: J. Biol. Chem. *247,* 4839 (1973)
167. Klug-Roth, D., Fridovich, I., Rabani, J.: J. Am. Chem. Soc. *95,* 2786 (1972)
168. Bannister, J. V., Bannister, W. H., Bray, R. C., Fielden, E. M., Roberts, P. B., Rotilio, G.: FEBS Lett. *32,* 303 (1973)
169. Pick, M., Rabani, J., Yost, F., Fridovich, I.: J. Am. Chem. Soc. *96,* 7329 (1974)
170. Lavelle, F., McAdam, M. E., Fielden, E. M., Roberts, P. B., Puget, K., Michelson, A. M.: Biochem. J. *161,* 3 (1977)
171. Bohnenkamp, W., Weser, U.: Biochim. Biophys. Acta *444,* 396 (1976)
172. Bettger, W. J., Savage, J. E., O'Dell, B. L.: Nutr. Rep. Int. *19,* 893 (1979)
173. Massie, H. R., Aiello, V. R., Iodice, A. A.: Mech. Ageing Dev. *10,* 93 (1979)
174. Gemsa, D., Leser, H. G., Seitz, M., Debatin, M., Bärlin, E., Deimann, W., Kramer, W.: Agents Actions Suppl. *11,* 93 (1982)
175. Bonta, I. L., Parnham, I. J.: Biochem. Pharmacol. *27,* 1611 (1978)

176. Lee, R. E., Lands, W. E. M.: Biochim. Biophys. Acta *260,* 203 (1972)
177. Maddox, I. S.: ibid. *306,* 74 (1973)
178. Vargaftig, B. B., Tranier, Y., Chignard, M.: Eur. J. Pharmacol. *33,* 19 (1975)
179. Boyle, E., Freeman, P. C., Goudie, A. C., Mangan, F. R., Thomson, M.: J. Pharm. Pharmacol. *28,* 865 (1976)
180. Letellier, P. R., Smith, W. L., Lands, W. E. M.: Prostaglandins *4,* 837 (1973)
181. Prohaska, J. R., Lukasewycz, O. A.: Science *213,* 559 (1981)
182. Vyas, D., Chandra, R. K.: Nutr. Res. *3,* 343 (1983)
183. Lukasewycz, O. A., Prohaska, J. R.: ibid. *3,* 335 (1983)
184. Rigas, D. A., Eqinitis-Rigas, C., Head, C.: Biochem. Biophys. Res. Commun. *88,* 373 (1979)
185. Lipsky, P. E., Ziff, M.: J. Immunol. *120,* 1006 (1978)
186. Brown, T. M., Bailey, J. S., Iden, I. I., Clark, H. W.: Inflammatory Diseases and Copper (ed. Sorenson, J. R. J.) p. 391, Clifton NJ, Humana Press 1982
187. Van der Goot, H., Pijper, P. J., Smit, H.: ibid., p. 409
188. Sorenson, J. R. J.: ibid., p. 289
189. Harris, E. D.: Proc. Natl. Acad. Sci. U.S.A. *73,* 371 (1976)
190. Ziche, M., Jones, J., Gullino, P. M.: J. Natl. Cancer Inst. *69,* 475 (1982)
191. Alessandri, G., Raju, K., Gullino, P. M.: Cancer Res. *43,* 1790 (1983)
192. McAuslan, B. R., Reilly, W.: Exp. Cell Res. *130,* 147 (1980)
193. Velo, G., Franco, L., Conforti, A., Milanino, R.: Inflammatory Diseases and Copper (ed. Sorenson, J. R. J.) p. 329, Clifton NJ, Humana Press 1982
194. Klevay, L. M.: ibid., p. 123

Human Insulin – Chemistry, Biological Characteristics and Clinical Use

Manfred Zoltobrocki and Rainer Obermeier

Hoechst AG, D-6230 Frankfurt/Main 80, FRG

Human insulin has recently become available for therapeutic use in diabetic patients. In this article, structure and preparation techniques, including isolation, total synthesis, semisynthesis as well as DNA-recombinant methods are discussed. Furthermore the biological properties in vitro and in vivo of human insulin are described. Finally, results from clinical studies in humans are reported and possible therapeutic advantages in comparison to animal insulins are discussed.

Progress in Clinical Biochemistry
and Medicine, Vol. 2

Introduction

The importance of insulin in the treatment of diabetes mellitus can be expressed in figures. Approximately 2 tonnes of insulin are used to treat diabetic patients throughout the world per year. This enormous figure gains in significance when expressed in terms of the daily average requirement per diabetic patient of 1.5 mg.

Porcine and bovine pancreases weigh about 40–80 and 250 g respectively. Using processes which have become more and more refined over the last 65 years, up to 140 mg insulin can be isolated from one kilogram of animal glands. A costly process of transporting the pancreases from the slaughterhouse to the insulin manufacturer in a deep frozen condition ensures that the tiny amounts of insulin in the pancreases are not cleaved and destroyed prematurely by the more than hundredfold surplus (by weight) of proteolytic enzymes which coexist in the pancreas.

Moreover, when one considers that up to 100,000 pancreases must be processed in order to obtain one kilogram of porcine insulin, it becomes clear why only animals intended for slaughter from large-scale livestock farming, i.e. pigs and cattle, can be used as a source.

Even under bovine or porcine insulin treatment, diabetes mellitus can be associated with a series of complications, such as nephropathy and retinopathy, cardiovascular damage and damage to peripheral blood vessels. It is generally believed today that the principle reason for these side effects is the unphysiological way in which the insulin is administered. This leads to unnatural fluctuations in blood sugar levels and the resulting peripheral lesions, although heterogeneity of animal insulins and differences in the amino composition, as compared to human insulin[1)] (Fig. 1) also have an effect, albeit limited, on the biological properties of the insulin. Naturally, it was thought that human insulin, as an endogenous hormone would fulfil the justified hopes and wishes of patients and doctors alike.

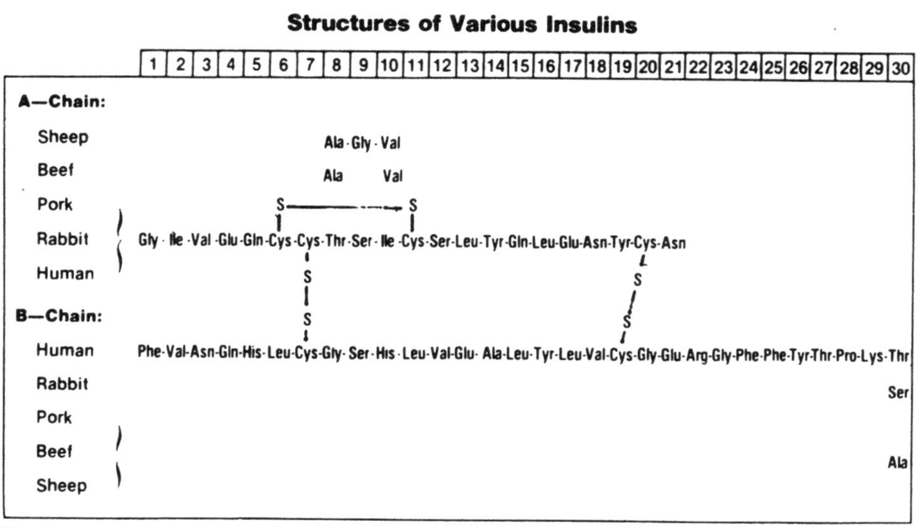

Fig. 1

A. Chemistry: Structure, Manufacture, Purification, Analysis

I. Structure and Characteristics

Human insulin – in the same way as insulins from other species – is secreted by the β-cells of the pancreas. Active insulin consists of two chains of peptides, A and B, which are connected by two disulfide bridges, an intra-chain disulfide bridge also being present in the A chain linking positions A6 and 11. The A chain is composed of 21 amino acids residues and the B chain of 30. During biosynthesis a linear precursor molecule, proinsulin, is first of all formed[11, 12], the C-terminus of its B chain linked to the amino terminus of the A chain by the C-peptide, which consists of 35 amino acid residues.

Human proinsulin is derived from preproinsulin, which is an even larger molecule. This precursor results as the first translation product in protein biosynthesis from the "insulin genes"[2–8]. The presequence of human preproinsulin is composed of a peptide consisting of 24 amino acids residues bound to the amino terminus of the B chain[2].

Contrary to proinsulin, preproinsulin is not obtained from pancreas tissue for analysis. The primary structures of the known insulin presequences have only been able to be determined using in vitro translation products of mRNA from pancreatic β-cells or by DNA sequencing combined with molecular biological cloning techniques.

The reason for this is the enzymatic cleaving of the presequence during transport of the total protein through the membrane of the endoplasmic reticulum into the free cell lumen. The signal peptide plays a decisive role in this transport mechanism[13, 14].

The further enzymatic conversion of proinsulin into free insulin and C-peptide occurs in the Golgi apparatus. Here the insulin is stored in rhombohedric granules until exocytosis takes place[15, 16] to protect it from further enzymatic degradation.

It must be mentioned here that the most prominent variations in the primary structure of the different species of insulin lie in the C-peptide section[17]. Although, for example, human and porcine insulin differ only in that threonine at position B30 is replaced by alanine in the latter, there are 17 positions in the C-peptides which differ, and additionally 2 deletions in porcine C-peptide[18].

The sequences of the A and B chains in primates, i.e. man and monkey, are identical. Human insulin has, however, a leucyl residue at position C7 in the C-peptide chain, whilst monkey proinsulin has a prolyl residue in this position[19, 20].

The amino acid sequences of the therapeutically important human and porcine insulins differ only slightly, the single variation being the threonine/alanine difference at position B30 (Fig. 1). As can be assumed from their completely identical biological activity, the tertiary protein structure of these insulins is also identical.

Human insulin, like porcine insulin, crystallises into rhombohedric 2 Zn (4 Zn) crystals, in which the insulin is present as three identical dimers forming a hexamer aggregate[21].

Three-dimensional X-ray structural analysis of 2 Zn crystals of human and porcine insulin at a resolution of 1.9 Å revealed marginal differences in conformation at position B30 and transposition of the crystal water molecules. Slight differences in the interaction between the amino acid side-chains were also established. These can be explained by the differences between threonine and alanine as far as extensibility and capability of forming hydrogen bridges are concerned. These small differences exert little or no influence on the structural elements responsible for receptor binding, aggregation and the way in which the insulins behave immunologically and chemically.

This only applies because no sequences are affected which have developed as invariant domains during the evolution of animal insulins[29)]. Recently, abnormal human insulins have been identified in serum and pancreas samples from diabetic patients. When secreted in an hyperinsulinaemic state, these insulins have led to typical hyperglycaemia of diabetes mellitus[23–26)]. The patients, however, react normally to exogenous insulin. So far, by means of HPLC, radioimmunoassay, immunoaffinity investigations, molecular cloning and DNA sequencing[27, 28)], 2 natural variants of human insulin have been found and elucidated. In one variant, human insulin B25 (Phe → Leu), leucine has replaced phenylalanine at B25[23, 24)]. A further variant is human insulin B24 (Phe → Ser)[25)]. The structures were confirmed by comparing them with corresponding semisynthetically obtained analogues[30, 31)]. When investigating the variant insulins B24 (Phe → Leu) and B25 (Phe → Ser) it was found that the insulins manipulated at position B24 still posessed approx. 15% of the potency of normal human insulin, as did insulin B24 (Phe → Leu). The biological potency of the B25 variant amounted, however, to only 2% of the natural activity[26)].

Interesting knowledge on conformative influences of these domains on biological activity was gained from these findings. The conformation of the B24 analogues is different, for example, this being shown by a reduction in the ability to dimerise[32, 33)]. B25 analogues retain the known insulin conformation. Losses in activity in both cases can be explained by the topography of the insulin molecule. The aromatic residue of B25 is on the surface of the insulin molecule and is directly involved in the receptor binding, whilst the aromatic ring B24 is directed into the molecule and stabilises the C-terminus of the B chain by van der Waals' interaction. As with B24 substitution, replacement of the aromatic substances with Leu or Ser indirectly or directly inhibits the hormone-receptor interaction.

First abnormal products of the insulin gene have also been found at the proinsulin level.

In some cases of hyperproinsulinaemia, a two-chain intermediate product of proinsulin has been identified in which C-peptide has already been cleaved from the B chain but is still bound to the N-terminus of the A chain[34, 35)]. The C-peptide in this intermediate product is linked by a Lys^{64}–Arg^{65} to the A chain[36)]. It is cleaved into insulin and C-peptide during the biosynthetic processing of normal proinsulin[36)]. Analytical investigation showed that in the abnormal intermediate product the second basic amino acid, $Arg^{65)}$, necessary for enzymatic recognition of the conversion point, is either missing or has been replaced by an amino acid residue yet to be determined.

The cases of natural mutants of human insulin and proinsulin found so far lead one to assume that in the future further natural variants will enable the discovery of structure and function relationships, which until now has been impossible because of the inaccessibility of the compounds.

With the exception of these few cases where the structure varies and which play a very small – if tragic – role in the enormous field of diabetes mellitus, the structural, chemical, physical and biological properties of human insulin are the same as those of bovine and porcine insulin in almost every respect[37,38]. The most important common feature is their biological activity i.e. the blood sugar lowering effect, which amounts to at least 26 U/mg[40] and to 28 U/mg for highly-purified human insulin, compared to the international insulin standard.

II. Production of Human Insulin

1. Natural Human Insulin

Human insulin can, of course, only be obtained from human pancreases for analytical purposes. Special processing procedures for small amounts of pancreas were developed for this purpose[29, 41–44]. Methods of isolating extremely small amounts of insulin were used for the elucidation of the structure of human insulin variants[26].

2. Chemical Synthesis

From the beginning of the 60s, much time and effort was put into the total synthesis of insulin. The oxidative recombination of reduced A and B chains[45, 46] paved the way to the first successful attempts at synthesising insulin[47–50]. Crystallisable, biologically fully-active bovine insulin was produced in this way in 1965[51]. Human insulin was obtained for the first time in 1967 from separately constructed chains[52]. This method of synthesis consisted of more than 200 individual steps, but the yield of only a few per cent on recombination of the separate chains was so low that it was inconceivable to produce human insulin in this way.

An interesting method of totally synthesising human insulin had been discovered by 1974[53]. This was a very valuable step indeed, as this method was specifically concerned with the synthesis of the disulfide bridges and, was the first to demonstrate the structure of insulin. Only 2 different S-protective groups were required for synthesis, since the starting material was an asymmetrical cystine-peptide, A(20–21) – B(17–20), which had already been prepared in 1973[54]:

```
           H-Cys-Asn-OBu^t  (Bu^t = tert.-butyl)
             |
Trt-Leu-Val-Cys-Gly-OH
```

Successive addition of the fragments Bpoc-A(14–19)–OH and H–B(21–30)–OBu^t resulted in an intermediate product which already possessed 22 of the 51 amino acids of insulin.

The asymmetric disulfide is apparently stable enough in this large molecule. All third functions of the amino acids involved are blocked by protective groups on a tertiary butyl basis. They can later be cleaved off acidolytically with trifluoroacetic acid. The Bpoc residue is an N-protective group[55], which can also be cleaved off acidolytically. This process is, however, so mild that tertiary butyl residues remain intact.

Further synthesis placed great demands on the selectivity of the protective groups. In order to continue at all, new conditions for synthesis had to be found, under which it would be possible to differentiate further between the already extremely acid-labile protective groups Bpoc (2-(4-biphenylyl)-propyl-(2)-oxicarbonyl-) and Trt (trityl-). A way was finally found by using pH-controlled acidolysis of the N-trityl and N-Bpoc groups with trifluoroethanol as solvent. A further obstacle was still present, however, in the shape of the synthesis of the A(1–13)–OH segment.

The cyclic disulfide Cys^6–Cys^{11} was attained by iodine-oxidation of the corresponding S-trityl cysteine residue. In doing so it was necessary to create conditions under which the S-acetamidomethyl group (S-Acm), also sensitive to iodine, would not yet react. This obstacle was also overcome and permitted construction of the entire protected sequence, still containing the cysteine pair A7–B7, protected by S-Acm: Cleaving off the tertiary butyl protective groups and iodine oxidation led the way to human insulin which was extensively purified by countercurrent distribution.

Although this method of synthesis first made gram amounts of human insulin available, as a production method it also proved to be much too expensive. Further methods of obtaining synthetic human insulin were based on findings made by Steiner[59] who, after reduction of the disulfide bridges and oxidative recombination, observed a recombination yield of approx. 70% with proinsulin. Laborious trials in which trypsin and carboxypeptidase B[61] were used to cleave human insulin enzymatically from synthetic proinsulin also showed that such a pathway was not suitable for production.

Recombination trials were also carried out in which instead of C-peptide the A and B chains of the insulin between the amino groups $N_{\alpha}A1$ and $N_{\varepsilon}B29$ were linked by dicarbonic acid. Since the distance between the two amino groups in the spatial structure amounts to only 8–10 Å, the fixation of these amino groups with a dicarbonic acid residue actually increased the recombination yield up to 70–75% of the theoretical value[62, 63].

The introduction of cleavable bridging reagents such as α,α'-diamino-suberic acid[64, 65], showed that it was possible to synthesise the correct disulfide bridges with linked A and B chains. Subsequent Edman degradation of the diamino suberic acid residue resulted in insulin[64].

All these attempts at chemical synthesis were, however, overtaken by developments in the fields of semisynthesis and genetic engineering, which were taking place at the same time. Nevertheless, these attempts were worthwhile, as they brought numerous new insulin analogues to light which were able to be investigated for structure and function[38, 56–58].

3. *Semisynthesis*

Since the only difference between human and porcine insulin is the presence of alanine instead of threonine at position B30 in the former, it seemed obvious to use semisynthetic procedures to convert porcine insulin into human insulin.

An experiment of this sort had already been published as early as 1972[66]. The six free carboxyl groups of porcine insulin had been converted into the corresponding methyl esters with diazomethane. The porcine insulin-hexamethyl ester was then converted by enzymatic cleavage with trypsin at Arg B22 and Lys B29 into desocta-peptide-B23–30-insulin-pentamethylester.

After protecting the N_αA1 and N_αB1 amino groups, which were still free, with the tert-butyloxicarbonyl residue, a molecule resulted which, in accordance with classic peptide chemistry, contained all carboxyl and amino groups in blocked form, with the exception of the carboxyl group at Arg B22, which was necessary for coupling with a synthetic octapeptide of the human insulin sequence B23–30.

After chemical peptide coupling, the protective groups should have been cleavable by acidolysis and basic conditions. Investigations carried out by other workers[67–69] revealed that the original yield of 75% human insulin was not reproducible. The reason for this was an almost quantitative imide formation at Asn A21 during the saponification of the methyl ester[70, 71]. Another chemical process for the conversion[69] of porcine insulin took these findings one step further by circumventing the protection of the carboxyl groups.

$N_\alpha A^1 N_\alpha$B1-Bis-tert-butyloxicarbonyl-desoctapeptide-(B23–30)-insulin was transformed by means of dicyclohexylcarbodiimide/N-hydroxybenzotriazol[72] and the N-terminal free, partially-protected human octapeptide. Despite the 6 free carboxyl groups, the –COOH residue of Arg B22 reacted preferentially. After cleavage of the protective groups, the actual problem was the chromatographic separation of so-called "isoinsulin" [69] from active human insulin. The small amounts of human insulin finally obtained were identical with the natural hormone.

The breakthrough for semisynthesis, was, however, when enzyme-catalyzed peptide-bonding procedures were applied which had been known for quite some time[73, 74].

As most insulins contain only one Arg B22 and one Lys B29, the C-terminal region of the B chain can be split off selectively by trypsin, which cleaves the peptide bonds specifically at the carboxyl site of basic amino acids. During the past few years a highly specific lysyl-endopeptidase[84] has been made commercially available, which even allows selective cuts between only Lys B29–Ala B30. The resulting des-Ala-B30 insulin (DAI) can be prepared alternatively by a very critical digest with carboxypeptidase A[85]. Degradation of the C-terminal Asn-A21 occurs, however, to some extent. After all desoctapeptide-B23–30-insulin (DOI) or DAI, two readily available starting materials, can be used for the semisynthetic exchange of the C-terminal sequence of the porcine insulin B chain. Starting with DOI, chemical coupling methods as mentioned result in poor yields and therefore extensive purification procedures. Substitution of the chemical coupling reagents with trypsin, however, raised yields to 60–80%[75]. Enzymatic coupling of DAI with threonine esters[77] was an interesting simplification, based on the observation that under certain conditions the cleavage of the Arg B22–Gly B23 bond by trypsin is suppressed. When lysyl-endopeptidase[84] became available, even this potential side reaction was able to be excluded[78].

Very recently, various groups[79–82] have independently found an essential improvement in the enzymatic semisynthesis of insulin. Under slightly variable conditions, single-step trypsin-catalyzed transamidation can be performed at the Lys B29–Ala B30 peptide bond.

The enzymatic conversion of porcine insulin into human insulin has been expanded in the last few years into production methods[80, 85], which differ only in the conditions of the reaction, such as pH values and application of alternative organic solvents or different threonine esters. Conversion rates of 80–90% are achieved. After splitting off the protective groups and purification, a total yield of 55–60% human insulin is achieved.

Ion exchangers or silica gel filtration or a combination of both are used for purification (see Sect. A. III.). These chromatographic procedures, and particularly purification on silica gel, ensure that unconverted porcine insulin and other pancreatic impurities such as proinsulin, somatostatin, pancreatic polypeptide, VIP, neurotensin, GIP and gastrin are eliminated[86]. The procedures are carried out at the level of the lipophilic human insulin ester, which too is favourable for purification using ion-exchange chromatography, since the negative charge is reduced by the ester function, compared to the unreacted porcine insulin.

Human insulin preparations manufactured using semisynthetic procedures are very pure. Practically no impurities are detected with modern, highly-sensitive methods such as HPLC and radioimmunoassay, even though their limits of detection are partly lower than 1 ppm. In addition to the production of therapeutically relevant amounts of human insulin, these semisynthetic methods have been used to prepare a series of insulin analogues with amino acid exchanges in the C-terminal region B23–30. By doing so, mutants of natural insulin[26] have been made accessible for assessment of structure and function (see Sect. A. I.). Unnatural B30 analogues have also been gained[85, 87] and some used for immunological investigations[87, 88]. These investigations showed that analogues of this type have the same degree of immunogenicity as human insulin, but that some differ considerably in their antigenicity towards preformed anti-insulin antibodies. Their behaviour is therefore similar to that of des-Phe B1 porcine insulin, which was investigated at an earlier date[89].

Consistent findings have yet to be reported on the mechanism of the single-step, enzymatically catalysed transamidation. Final conclusions cannot be drawn on the positive[85] and negative[90] findings. Clarification of this question will have to be the subject of future studies.

4. Biosynthesis of Human Insulin by Means of Genetic Engineering

As with the development of peptide and protein chemistry, the quest for human insulin was the driving force behind the development of genetic engineering techniques in this field.

Two ways are known of obtaining human insulin from modified E. coli by fermentation. They both adhere strictly to the historical development of insulin synthesis. The first method developed was preparation via separately biosynthesised A and B chains, this was followed by the biosynthesis of human proinsulin, from which human insulin can be derived by enzymatic cleavage[61], as mentioned earlier.

In both cases the insulin sequences first of all occur bound to an E. coli specific protein and are isolated in this form.

In the case of the separate chains, chemically synthesised genes[96] are inserted into a plasmid, pBr322[95], which contains E. coli sequences of the lac-operon and in this way control the expression of the inserted DNA. Protein synthesis, therefore, results in a chimeric protein which consists of a β-galactosidase-insulin chain[93]. The fusion proteins are isolated, purified and further processed into human insulin using known protein-chemical methods. For this purpose, the fusion proteins were constructed in such a way that a methionine residue cleavable by cyanogenbromide was inserted between the β-galactosidase and the insulin chain. After purification, sulfitolysis was performed following cleavage with cyanogenbromide and the resulting S-sulfonates of the insulin chains formed the classic starting material for the recombination of the disulfide bridges into insulin as described at A. II. By varying the latter method in decades of research, scientists finally managed to increase the yield of the S–S-bond to such an extent[93] that it was possible to develop a productive manufacturing process for human insulin.

It has been possible to improve the fermentative production of human insulin by changing to proinsulin instead of using separate chains, the very important improvement being, however, the replacement of the lac-operon with the trp system. The chimeric β-gal-protein consists of 1,004 amino acid residues, but the trp system uses only 191[92]. This means that the content of insulin in the trp-fusion-protein described is five times higher (86 amino acid residues for proinsulin) than with β-gal-fusion[92]. In addition a manyfold higher expression rate has been observed for the trp-fusion[92]. The most significant characteristic of both expressed polypeptides is that they are very poorly soluble, occurring in the form of insoluble aggregates in the E. coli cell. These granules, which electron microscopy revealed to be strikingly large, protect the product from immediate enzymatic degradation by the E. coli proteases[99], of which at least two have been made responsible for efficient proteolysis. Direct expression of human insulin is, however, still impossible because of proteolytic degradation. Molecular biological investigations in the future will therefore definitely be concerned with overcoming the problem of proteolytic degradation by studying other microorganisms such as yeasts or Bacillus subtilis or modified intracellular transport mechanisms. Important for the present is, however, that the aim of producing insulin for therapeutic purposes has been reached by this breakthrough using pioneering genetic engineering methods.

So-called biosynthetic human insulin has been commercially available since 1982, as has the semisynthetic form.

III. Purification and Analytics

1. Methods of Purification

As far as the purification of human insulin is concerned, much of the experience made with insulins extracted from animal glands was able to be used[37, 100]. Ion-exchange chromatography with anion exchangers has been used here very successfully for the latter. The method using medium pressure liquid chromatography on silica gel with a mixture of organic eluents, mentioned in Sect. A. II. 3. was specially developed for the fractionation of insulin and insulin esters. The high separation performance described is a result of differences in lipophilic character between both compounds. The insulin esters carrying t-butyl groups are highly lipophilic which leads to baseline fractionation. The method is rapid, works with large loads and has the advantage that if the silica gel becomes contaminated it can be replaced cheaply. Method-immanent losses, such as protein adsorption amount to only a few per cent and are therefore negligible. The system has proved its worth as a large-scale production method.

2. *Analytical Methods*

The introduction of semisynthetic and biosynthetic insulins for therapeutic use raised both the sensitivity and range of methods of analysis to higher standards than ever before. Protein-chemical work on the insulin molecule necessitated demonstration of the structural identity and the absence of reaction by-products for both human insulin preparations. The research into the by-products meant that in extreme cases the analytical methods had to be sensitive enough to detect individual molecules (see Ref. 80, p. 42). In addition to the classic methods of amino acid analysis, DISC-electrophoresis, determination of biological activity, UV spectroscopy and radioimmunological and radioreceptor measurements, reverse-phase high pressure liquid chromatography has proved to be an extremely sensitive method of determination[79, 92, 101, 102]. The outstanding performance of RP-HPLC is demonstrated by the way in which it clearly fractionates bovine, human, porcine and mouse[101] insulins out of mixed samples. Insulin esters, proinsulin, hydrolytic and enzymatic products of degradation and chemical derivatives of insulin[92] can be demonstrated equally as distinctly. This means that a method of determining the byproducts of human insulin production is available which will detect amounts as small as 100 ppm. With regard to the demonstration of the correct disulfide linkage when recombining separate insulin or pro-insulin chains, reliable and quantifiable results have recently been reported using HPLC fingerprinting of S. aureus protease V 8 fragments of the various insulins[103, 104]. Thanks to its wide spectrum of applications, sensitivity and reliability, HPLC has become the work-horse of insulin quality control investigations[39].

A range of highly specific analytical methods is therefore available today, complemented, in the case of biosynthetic human insulin, by radioimmunoassay[15] for the detection of E. coli fragments, and the limulus-amoebocyte lysate assay[106] for the determination of the endotoxin content.

B. Biological Characteristics and Clinical Use in Man

I. *Introduction*

On 30th July 1921 in Toronto, Banting and Best injected a dog with an active pancreas extract for the first time. This was the end of a long era in which the diagnosis "diabetes mellitus" was almost synonymous with a death sentence for those affected. But this was also the beginning of a new era in which man would learn how to use insulin as a therapeutic tool.

Today, more than 60 years later, man is still learning. In the early days of insulin treatment the most important objective was to purify the raw extracts – then full of impurities – to such an extent that the most severe (mostly allergic) side effects no longer occured. Later, the emphasis shifted to the development of longer-acting insulin preparations. Several injections were needed per day, and the aim was to reduce this number considerably. Recently, advances have been made in both the areas of purification and product development.

It was not until man had been treated for many years with insulins of animal origin that close attention was paid to the structure of the insulin molecule. The amino acid sequence of insulin was elucidated by F. Sanger in the fifties. Subsequent research revealed that the insulin molecules of various species are made up of a number of amino acids which vary (29 positions) and a number of amino acids which are fixed (22 positions)[29]. Human insulin differs from known animal insulins in a number of different amino acids. The blood-sugar-lowering effects of the different insulins are, however, identical in most cases.

Human insulin has come to be used for treatment only recently, a not unimportant factor in encouraging its use being the explosive developments in the area of genetic engineering.

The following important questions emerged during its development:

1. Are the effects of human insulin in vitro and in animal models comparable with the effects of animal insulin?
2. Are the biological effects of single doses of human insulin in healthy subjects comparable with those of animal insulins?
3. Do the effects of human insulin differ from those of animal insulin after single dose and long term treatment in diabetic patients?
4. Does human insulin have therapeutic advantages and what are its indications?

These questions are discussed in the following.

Differentiating between the insulins by manufacturing process seemed inappropriate, particularly since all investigations, including X-ray structural analysis[107], brought no structural differences between semisynthetic and biosynthetic insulin to light. Only differences in the purity of the insulins might be present, and these are theoretical. Although impurities may be responsible for side effects, modern, highly-developed purification techniques mean that this is of no practical importance.

II. Biological Characteristics

1. In Vitro Effects on Cells and Organs

Binding to specific receptors on the surfaces of cells is the first step in the mode of action of insulin. This includes binding to cells which have a non-insulin-dependent metabolism. The red blood cells (erythrocytes) are an example of these. Despite this, the binding of insulins with different structures to cells of this type and others has been compared, since this constitutes a simple experimental model.

Investigations have been carried out on the binding affinity of human insulin – where appropriate compared with animal insulins – to erythrocytes, monocytes, adipocytes, hepatocytes, lymphocytes and round cells of different origin. Changes in binding affinity and the number of receptors per cell have also been investigated after switching patients to human insulin.

a) Blood Cells

The binding of the different insulins to the cells circulating in the blood is usually investigated with radiolabelled insulin. In addition to the insulin structure, the position of the iodine label is important (the label can be attached to positions A-14 (tyr) or A-19 (tyr) or both).

Binding of human insulin to human erythrocytes has been investigated in healthy subjects and diabetic patients and compared with porcine insulin. The results differed in both healthy subjects and diabetic patients. In three studies[108–110] no differences emerged between human and porcine insulin. In a fourth, however, the binding affinity of the porcine insulin was markedly lower than that of the human insulin at low insulin concentrations[111].

Binding to human monocytes has also been investigated in healthy subjects and diabetic patients. No differences emerged between human and porcine insulin in healthy subjects[109, 112, 113]. Binding of human insulin was found to be stronger than of human proinsulin[114]. As with healthy subjects there was no difference in the binding of human insulin and porcine insulin to monocytes in diabetic patients[112]. Finally, no differences emerged between the binding of human and porcine insulin to human lymphocytes (IM-9) in vitro[110, 115–118].

The picture was slightly different for the degree to which human and porcine insulin bind to the blood cells of diabetic patients during long term treatment. Raised receptor affinity (erythrocytes) was discovered after changing to human insulin in newly-diagnosed Type I and Type II diabetic patients[119] and children[120]. However, the concentration of receptors in the children then decreased[120]. The binding of insulin to monocytes did not appear to differ when newly-diagnosed diabetic patients were treated with human insulin or porcine insulin[121]. Investigations of rheological variables revealed that the effects of human insulin on blood and plasma viscosity, haematocrit, erythrocyte sedimentation rate and deformability were not different from those of animal insulin[122].

b) Other Cells and Organs

Binding studies on rat adipocytes[110, 117, 118, 123–126] showed no differences between insulins from different species. There were also no differences as far as glucose transport[124, 127], lipogenesis[125], lipolysis[117] and ATP depletion[117] were concerned. Binding to rat hepatocytes, their isolated membranes[128] and pig hepatocytes[129] was not different for human and animal insulins, nor was the influx of aminobutyric acid. Rat epididymal fatty tissue[117], the soleus muscle of the mouse[128] and rat islet cells[130] did not react differently to human and porcine insulin. The same effects on the rat diaphragm were observed with human insulin and porcine insulin[126, 131]. The rat fatty cell assay revealed higher activity for porcine insulin[131]. No differences emerged for human round cells[132]. Human insulin and sperm whale insulin were compared on rat adipocytes and chicken embryo fibroblasts[127]. The sperm whale insulin was more active as far as glucose transport into the cell was concerned.

No further differences were observed.

2. *In Vivo Effects in Animals*

a) Kinetics

The decrease in radioactivity from the circulation was investigated in rats after i. v. injection of radiolabelled insulin[134]. Products of degradation were chromatographically separated before determination. The elimination kinetics of human and porcine insulin proved to be similar. The respective metabolic clearance rates (MCR) did, however, differ: at 24.6 ml/min · kg the value for human insulin was higher than that for porcine insulin at 20.8 ml/min · kg.

b) Effects on the Blood Sugar

Single dose toxicity and general pharmacological properties were the same for human and porcine insulin[135, 136]. Establishing whether human insulin differs from animal insulin in its effects in animals is of little relevance to the treatment of diabetic patients. Such investigations are, however, unavoidable as part of a preliminary testing and safety programme (special pharmacology). The rabbit test resulted in a value of 27 U/mg for the blood-sugar-lowering effects of human insulin[115]. This was the same as the value for porcine insulin. No differences emerged in the mouse convulsion test[126]. The rat i. v. test showed that human insulin was more potent than porcine insulin[134]. This test is, however, less reliable than the others.

III. *Clinical Use in Man*

1. *Healthy Subjects*

a) Kinetics

Many of the kinetic findings reported below can only be subject to limited comparison. The comments are often based on historic comparisons i. e. results from different studies and in other cases the insulin dose and mode of application (s. c., i. v. or other) were different.

Of interest were the course of insulin levels after injection of different insulins and in particular the maximum insulin level, the time it occurred and the total amount of insulin absorbed, which is indicated by the area under the insulin/time curve (AUC).

Insulin levels after subcutaneous injection of different doses of human insulin have been compared with those after other insulins. The insulins have also been tested in different preparations such as short and long-acting. Fast and short-acting regular insulins were compared at the following doses after subcutaneous injection: 0.05 U/kg[137, 138]; 0.075 U/kg[139, 140, 144]; 0.1 U/kg[137, 142–145]; 0.15 U/kg[132, 144]; 0.3 U/kg[146, 147]; 10 U[148, 149]; 4.8 U and 9.6 U[150] were given regardless of bodyweight.

On occasions, human insulins from different manufacturers have been compared. However, little can be concluded from these investigations as they fall into the category of the above mentioned historical comparisons. Identical or similar porcine insulin preparations were usually compared with human insulin. Although when viewed as a whole, the results are contradictory, a marked trend towards higher serum levels with human insulin emerged.

Higher maximum concentrations with human insulin emerged after administration of the highest dose of 0.3 U/kg[146, 147]. The AUC value was greater for human insulin than for porcine insulin after 0.05 U/kg[137, 138]; the same applied to 0.01 U/kg[143]. At the same dose, however, exactly the opposite was found by other investigators[144, 145] and although a third found higher levels after human insulin with 0.05 U/kg, he found no difference after 0,1 U/kg[137]. With somatostatin administration to suppress endogenous insulin secretion, no difference emerged for 0.075 U/kg[139, 140]. After 0.15 U/kg both identical insulin levels[132] and higher levels with human insulin were observed[144].

After injection of fixed doses per subject, higher levels were observed for 10 U and 9.6 U of human insulin[148, 150], but not after 4.8 U[150]. Adding a protease inhibitor (Trasylol) did not affect insulin levels[142]. This means that possibly more slight proteolytic degradation is not responsible for the occasionally higher levels after human insulin. Differences in the courses of insulin levels after subcutaneous injections of fast and short-acting human and porcine insulin were observed only after 10 U: peak concentrations of human insulin were reached after 110 min as compared to 68 min after porcine insulin. In the same study, however, the t_{max} values for a human and a porcine insulin from two other manufacturers were the same. Here the values were 120 min (human insulin) and 118 min (porcine insulin) and 70 and 84 min respectively[149].

Similar investigations were carried out with subcutaneously injected long-acting insulin preparations. No differences between AUCs were observed after 20 U and 0.25 U/kg zinc suspensions of human and porcine insulin[144, 148, 149]. There were also no differences worth mentioning with preparations with different delayed release mechanisms such as the NPH form[144, 149, 151]. In some studies, however, higher levels with human insulin were again observed[152, 153].

Insulin levels after intravenous injection of regular insulins (human insulin and comparative preparations) were also investigated after different doses: 0.025 U/kg[137, 138]; 0.03 U/kg[143]; 0.05 U/kg[137, 138, 154]; 0.075 U/kg[141, 155]; 0.1 U/kg[132, 144, 154–158]; 0.15 U/kg[144, 159].

Differences in kinetics between human and porcine insulin were found only for AUC values after intravenous injection of 0.1 U/kg[154, 157]. The area for human insulin was greater. The readings and calculated variables did not differ at all other doses.

In addition to intravenous injections, insulin was also given by infusion. This method enables the metabolic clearance rate (MCR) to be calculated. The infusion rates differed: 25–196 mU/min[160, 161]; 16.7 mU/min and 28.3 mU/min[150]; 20 mU/kg · hr[162–164]; 32 mU/kg · hr[163]; 50 mU/kg · hr[162–164]; 20 mU/m² · min[165].

In some cases the so-called glucose clamp procedure was used[159, 162–164]. With this method the blood sugar is kept almost constant by using the Biostator, a blood-glucose-controlled infusion system for continous insulin and glucose infusion. The metabolic clearance rate, terminal half-life and volume of distribution were the same for human and porcine insulin[159, 160, 162, 164, 166]. There were no differences between the insulin levels reached after infusions of human and porcine insulin[150, 163].

b) Effects on Blood Sugar and Other Variables

α) Blood Sugar

Although findings made on the effects of insulin preparations in healthy subjects cannot be applied directly to diabetic patients, they nevertheless provide important information comparing different insulin preparations. As for the kinetic data, historical comparisons between different studies are of only limited value.

A large number of studies have been carried out on the efficacy of various human insulins. The aim of these was to establish whether the slight structural differences from other species of insulin bring about different hormonal effects.

As with the determination of insulin kinetics, the different routes of administration (s. c. injection, i. v. injection and i. v. infusion), different doses and different galenic preparations must be taken into account. Regular insulin (short and fast-acting) was investigated at the following doses: 0.05 U/kg[137, 138]; 0.075 U/kg[139–141]; 0.1 U/kg[137, 138, 142, 143, 145, 168, 169]; 0.15 U/kg[132, 169]; 0.3 U/kg[147]; 4.8 U/subject[150, 170, 171]; 9.6 U/subject[150, 170, 171] and 10 U/subject[148, 149, 172].

In most cases porcine insulin used for comparison was either galenically identical or presented in a similar form.

Few differences in effect emerged:
The blood sugar lowering effect of human insulin was more marked after 10 U/subject[148]. This was also the case after 0.1 U/kg[143] and 0.05 U/kg[138]. With Biostator counterregulation at the higher dose of 0.3 U/kg, the amount of glucose required to compensate for the effects of the insulin was higher with human insulin. The onset of effect was also more rapid[147]. When two doses were compared, at the lower dose the human insulin was found to be slightly more effective. At the higher dose, however, the human insulin was found to be less effective or equally effective[138, 171].

The majority of investigators found no differences between human insulin and the comparative insulins as far as the blood sugar lowering effect in healthy subjects was concerned[132, 139, 140, 145, 149, 150, 168, 170, 172]. It is interesting to note that occasionally higher insulin levels were obviously not necessarily connected with lower blood sugar levels.

The addition of a protease inhibitor to the injection solution did not result in any difference in pharmacodynamic effects[142].

The situation is different with intravenous injections:
In this case the rate of diffusion from the site of injection is hardly affected by the physico-chemical properties of the insulin molecule. Differences occurring after i. v.

injection should, therefore, be attributable to intrinsic differences in effect between human and comparative insulins.

The following doses were given: 0.01 U/kg[174]; 0.025 U/kg[138, 162]; 0.03 U/kg[143]; 0.05 U/kg[137, 138, 154]; 0.075 U/kg[141, 155, 174–176]; 0.1 U/kg[132, 154–158, 168, 169, 174, 177, 178]; 0.15 U/kg[132, 159, 168]. As after subcutaneous injection, differences between human and animal insulin were found only rarely.

In two studies, the effects of human insulin were found to be more marked than those of porcine insulin after 0.1 U/kg[158, 174]. In a further study, however, the reverse was the case: the blood sugar lowering effects of porcine insulin were more marked after 0.05 U/kg[154]. Most investigations revealed no differences, regardless of the dose[132, 137, 138, 143, 150, 154–159, 168–170, 174–178]. The well established, more rapid and more marked effects of shorter duration after intramuscular as opposed to subcutaneous injection were also observed with human insulin[141].

Glucose-clamp trials with infusion of various insulins were performed to investigate the biological activity of the test preparations. Different infusion speeds were used: 25 mU/min, 49 mU/min and 196 mU/min[160]; 20 mU/kg · hr[162–164]; 32 mU/kg · hr[163]; 50 mU/kg · hr[162–164]; 1 U/hr[150, 170, 171]; 1.7 U/hr[150, 170, 171]; 2.4 U/hr, 7.4 U/hr[181] and 20 mU/m^2 · min[182].

The euglycaemic clamp was used in most of these investigations (blood sugar held as far as possible within the normal range). This eliminates the influence of the decreasing blood sugar levels caused by the infusion of insulin. This model revealed no definitive differences between human and porcine insulin[150, 160, 162–166, 170, 171, 181].

The same technique was used to establish whether the action of human insulin is different in various groups of healthy subjects and patients (such as obese and acromegalic patients). The effects were the same as those known with animal insulins[104].

The nature of extended release and long-acting human insulin suspensions is such that they were only given subcutaneously. Most of these were NPH formulations[146, 151–153, 168, 169] and zinc suspensions[148, 184].

Using the counter-regulatory biostator model NPH human insulin proved to be more effective than identically prepared porcine insulin (0.3 U/kg)[146]. The same applied to mixtures of regular and NPH insulin (0.4 U/kg)[152]. Finally, in one study, the effects of NPH human insulin were stronger only over a limited period after injection (5–11 h) at doses of 0.15 U/kg[151].

All other investigations[151, 153, 168, 169] with NPH formulations and zinc suspensions revealed no differences[148] or only very short-term differences[184] in the effects of delayed action insulins on blood sugar levels in healthy subjects.

β) Other Variables

In addition to lowering the blood sugar level, insulin brings about changes in a number of other biochemical variables, some resulting from the lower blood sugar level and others being a direct response to the effects of the insulin.

The effects of human insulin on the following have been investigated in healthy subjects: metabolic variables: lactate, pyruvate, free fatty acids, β-hydroxy-butyrate, and other ketones, glycerol, amino acids (single-chain and branched); counter-regulatory hormones: dopamine, adrenaline, noradrenaline, cortisone, growth hormone, glucagon, gastrin and prolactin; and variables of the electrolytes: potassium, phosphate. The suppression of endogenous insulin secretion by the administration of exogenous insulin has also been tested using C-peptide levels.

Very soon after testing with human insulin began, the symptoms of hypoglycaemia associated with low blood sugar levels appeared to be less marked than with porcine insulin. Catecholamines and glucagon are responsible for the rapid counter-regulation when hypoglycaemia occurs[133]. Comparative studies were therefore carried out to determine the amounts of counter-regulatory hormones released. Insulin was injected intravenously in most trials. The following doses were given: 0.01 U/kg[174]; 0.025 U/kg[137]; 0.03 U/kg[143]; 0.05 U/kg[137, 154]; 0.075 U/kg[155, 174, 176]; 0.1 U/kg[154–158, 176, 177] and 0.15 U/kg[159].

The same effects after subcutaneous injection were investigated in two studies[137, 145].

After 0.075 U/kg and 0.1 U/kg adrenaline levels were slightly lower under human insulin[155, 156, 174, 176, 178]. The symptoms were also less marked after human insulin[155, 174]. The same authors were, however, unable to reproduce these results[167, 173]. No or only minimal differences in the adrenaline response were found after 0.025 U/kg, 0.05 U/kg and 0.15 U/kg[137, 159]. Noradrenaline and dopamine showed either no change in the hypoglycaemia test (dopamine:[155, 174, 178]) or the response to human insulin and porcine insulin was the same (noradrenaline:[137, 155, 159, 178]). In three studies the glucagon response was identical after human and porcine insulin[137, 159, 177] and in a further two was slightly less marked under human insulin[145, 176]. In a second counter-regulatory phase, cortisone is released later than the catecholamines and glucagon[133]. Here also, some workers found no differences between human and porcine insulin[137, 143, 159, 177], some found that less cortisone was released with human insulin[137, 154–158] and others that cortisone release was higher with human insulin[154].

Similarily, the findings for growth hormone were also contradictory: decreased release after human insulin as opposed to porcine insulin[154–158] compared with identical amounts released[137, 143, 154, 159, 177] and even increased release in one study[176].

One worker reported that prolactin levels rose with porcine but not with human insulin[176] and identical increases were reported by another[159]. Gastrin appeared to behave the same after both[154].

Findings on the counter-regulatory systems in healthy subjects may not be applied to diabetic patients, since it is known that these complex systems are often disturbed in the latter[179].

The antilipolytic effects of human insulin were compared with those of porcine insulin by determining lipids in the blood. These proved to be more marked[145, 175, 176] or the same[163–165, 180]. One worker reported no change in triglycerides after intravenous, subcutaneous and intramuscular injections of 0.075 U/kg human insulin[141]. With measurement of ketones in the blood (3-hydroxy-butyrate and others), the antiketogenic effects of human insulin were found to be more marked[175, 176] or the same[140, 163, 164].

Human and porcine insulin affected intermediate products of the glucose metabolism and the amino acid metabolism in the same way[163–165, 175].

As far as the electrolytes are concerned, changes in potassium levels during insulin-induced hypoglycaemia are of interest since decreases in these levels clearly depend not only on the blood sugar levels, but are also affected by adrenalin[182, 185]. A number of investigations showed a more marked decrease in serum potassium after porcine insulin[174, 178] whilst others showed no difference[175, 176].

Finally, a number of workers investigated possible differences between the various exogenous insulins in the degree to which they suppress the secretion of endogenous insulin. This can be demonstrated by determining C-peptide, a cleavage product in the biosynthesis of insulin. No differences were found after intravenous infusion and subcutaneous injection of different amounts of insulin[138, 140, 143, 150, 152, 153, 155, 158, 159, 163, 165].

After injection of 0.05 U/kg of human and porcine insulin, again no differences between C-peptide levels were observed[145, 154]. In the same study, however, the doses were doubled and the human insulin was found to have a greater suppressive effect. Finally in two studies, weaker suppressive effects of human insulin were found after 0.1 U/kg i. v.[155, 157] which contradicts the findings of[154].

There must be many reasons for the contradictions in the results. For example, it was demonstrated that although human insulin inhibits the secretion of endogenous insulin in persons of normal weight, this is not the case with the overweight[186]. Other discrepancies may have been caused by different methods of investigation and analysis.

2. *Diabetic Patients*

a) Kinetics

Special techniques must be used to investigate the pharmacokinetics of insulin preparations in insulin-treated diabetic patients.

The insulin concentrations in the blood and their determination are affected by the following factors:

- circulating antibodies to insulin (IgG)
- residual endogenous insulin secretion (especially in Typ II diabetics)

Circulating antibodies to insulin may bind part of the exogenous insulin and in doing so decrease the amount of "free" insulin available. In addition to this, antibodies compete with the antibodies used in radioimmunological methods of determination and distort the results. In general, therefore, when carrying out pharmacokinetic investigations in diabetic patients, the insulin antibodies – sometimes with the insulin they have bound – are separated off and only the "free" insulin concentration is determined[187]. Residual endogenous insulin secretion is usually negligible or is eliminated mathematically.

The trend to higher insulin levels after human insulin observed in healthy subjects was practically never seen in diabetic patients. No differences in the concentration of free insulin and the course of insulin levels emerged when human and animal insulin were compared in subcutaneous injections of 0.2 U/kg[162] and 0.1 U/kg[145] and individual or fixed doses of 6 U and 24 U per patient[188–191]. This applied to short and fast acting regular insulins and insulin preparations of gradual onset with intermediate and long durations of effect. The insulin curves after infusion of the human and animal insulins did not differ[192–195]. A kinetic metabolism model for human insulin was postulated as early as 1973[161].

Human insulin resulted in higher serum insulin levels in only one investigation[196], after 0.15 U/kg subcutaneously. The comparative insulin in this case was, however, des-phe-insulin which leads to lower values because of the way its lower antigenicity affects the RIA[197].

Conversely, NPH porcine insulin resulted in higher levels than NPH human insulin[198].

b) Effects on Blood Sugar and Other Variables

α) Blood Sugar

The quantitative measurement of the most obvious and analytically easy to demonstrate metabolic effect of the insulin – the effects on the blood sugar – is difficult in the diabetic patient. In addition to the obvious effects of the dose given, other factors must always be considered in healthy subjects and are of particular importance in diabetic patients. The sensitivity to insulin of the latter is also affected by antibodies to insulin which may be present or by an altered receptor state (caused by obesity, for example). Baseline blood sugar levels are also important in clinical comparisons. Any conclusion drawn on the comparative efficacy of human and animal insulins in such cases will therefore always carry an element of uncertainty.

As in healthy subjects, a very large number of investigations on this subject have already been carried out. Different routes of administration, galenic formulations and of course different trial designs were used. A basic difference between single-dose studies and long term treatment studies must be made.

The biological activity of totally synthetic human insulin in diabetic patients was confirmed in 1979[232a]. Single-dose studies were carried out with subcutaneous injections of regular insulin. Control of the counter-regulatory glucose requirement with the Biostator revealed no differences between human, bovine and porcine insulin[162]. With simultaneous infusion of arginine, the effects of human insulin appeared to be more marked than those of porcine insulin for 30 min after subcutaneous injection of 0.2 U/kg. The reverse, however, was the case between 60 and 90 min after injection[145]. Intravenous infusion – particularly when combined with the Biostator – offers good controlled conditions for the comparison of regular insulins. Occasionally, the blood sugar is held in a particular range using the glucose clamp. This enables conclusions to be drawn on the biological activity of the insulin in question, based on the consumption of glucose. However, even with such studies, comparability is impaired because not always the same blood sugar range is chosen.

There was no difference in activity between human and bovine insulin (glucose clamp at 120 mg/100 ml, 2 infusion rates)[192]. No differences also emerged when human and porcine insulin were compared using the euglycaemic clamp and 4 infusion rates of insulin[159, 193, 194]. The same was found in Biostator trials without the glucose clamp[154, 199–203].

Regular insulin is not often used alone for therapy. It is, however, indicated in diabetic coma and for short term use when first stabilising patients on insulin. It is very difficult to carry out comparative studies in these indications and they can only be performed interindividually. Human insulin proved to be effective in these conditions and no differences from animal insulin were observed[204–208].

Recently, regular insulin has been used for the treatment of diabetes in portable infusion devices (pumps). There were also no differences between human and porcine insulin when given this way[148, 194, 209]. When used in pumps the insulin is subject to abnormal thermomechanical strain, which may lead to denaturation. Specially stabilised human pump insulins have been developed to avoid this, and used for therapy[210, 211].

Long-acting insulins are presented as suspensions and are therefore only suitable for subcutaneous administration. Single dose studies showed no great differences from porcine[193, 212] and bovine insulins[213]. Porcine insulin even emerged as more effective in a glucose-clamp study[198].

Most studies are therapeutic studies of different duration. Human insulin in different preparations was compared with different preparations of animal insulin, in some case interindividually, in others intraindividually. In doing so, a distinction had to be made between newly-diagnosed patients and those being switched from other insulins. The main conclusion which can be drawn on these studies is the demonstration of efficacy and tolerance of human insulin in long-term treatment[119, 120, 168, 180, 190, 193, 213–232a]. This was also demonstrated for diabetic children[204, 233–236]. Human insulin suppresses hepatic production of glucose in the same way as porcine insulin[194].

Occasional slight differences occurring under study conditions cannot always be fully explained. In some cases they can be attributed to the way in which the study was conducted, rather than the insulin preparation in question. Higher mean blood glucose (MBG) levels under therapy with human insulin than with porcine and bovine insulin were observed[220], just as higher levels were also observed during the night[188] and fasting in children[233]. The authors concluded that the delayed-action human insulins used had shorter durations of effect[234, 237].

Human insulin sometimes had to be given in higher doses than porcine insulin, but not bovine insulin[220]. However, lower doses of human insulin were also sometimes required[235]. Some workers reported that HbA_{1c} (glycosylated hemoglobin) levels were lower after human insulin[180, 214, 230] and other the blood sugar levels[190, 200, 212, 219, 222, 226]. In some of these cases, however, only single readings from frequent and regular sampling schedules were compared and in others the previous treatment was used a reference.

The onset of the remission phase was reportedly also earlier and its intensity better under human insulin[236].

However, in most cases, no differences between human insulin and animal insulins were observed in diabetic patients.

It is interesting to note here that human insulin was used to treat a vegetarian diabetic successfully. He had previously refused treatment with animal insulins[238].

β) Other Variables

As in the healthy subjects, the effects of human insulin on metabolic variables other than glucose were also determined in diabetic patients.

The question of counter-regulation in hypoglycaemia caused by human insulin has already been discussed for the healthy subjects. The difficulty in applying such findings to the diabetic patients was also mentioned. The counter-regulatory hormones were also measured in diabetic patients. Glucagon levels under human insulin were the same as under bovine[192] and porcine insulin[216, 227]. Under arginine infusion, glucagon levels were lower after human than after porcine insulin[145]. No differences between human and bovine insulin were established as far as cortisone levels were concerned[192]. The secretion of growth hormone under human insulin was investigated, but not compared with other insulins. It is interesting to note that it had already been secreted before the blood sugar level was really low[239]. Variables of the antilipolytic and antiketogenic effects of human and animal insulins were identical[162, 188, 192, 194, 213, 227] as were HDL-cholesterol levels[231]. There were also no differences in the effects on the amino acid metabolism[162, 194].

c) Undesired Effects

α) Immunogenicity

One of the great hopes during the development of human insulin was concerned that the level of immunogenicity would be lower, i. e. that fewer specific antibodies to human insulin would form under therapy with exogenous human insulin. It was even hoped that treatment with homologous insulins might not lead to the formation of any antibodies. It has, however, become apparent that this hope has not been fulfilled. The immunogenicity of insulin is determined only in part by its structure. Although it is known that bovine insulin, with three different amino acids as compared to human insulin (A-8, A-10 and B-30), has a higher degree of immunogenicity than porcine insulin, with only one position different from human insulin (B-30), it must be kept in mind that immunogenicity of insulin is also determined by other factors such as the mode of preparation, the route of administration, the dose and the time it is given. In addition the formation of antibodies is subject to genetic control. HLA-DR 4 indicates a "high-responder" rating[240].

Human insulin proved to be less immunogenic in mice than porcine and bovine insulin[241]. The relevance of this finding to man must remain open.

Human insulin was found to be immunogenic when used to treat diabetic patients. Only very few controlled studies have been carried out with regard to this. The nature of the subject matter is such that only interindividual investigations can be carried out. As far as new insulin patients are concerned, the percentage of patients developing IgG antibodies after human insulin within a certain period of time was smaller than under porcine and bovine insulin: after 6 months, 14% as opposed to 29%[112, 240] and 35% as opposed to 61%[242]. In other studies IgG antibodies had formed in all patients after 3 months' treatment with human and porcine insulin[243]. After 12 months in another study, 55% of the patients on human insulin and only 40% on porcine insulin were clear of insulin-specific IgG[244]. Similar findings were made in new paediatric insulin patients[245, 246]. Antibody titres also proved to be lower under treatment with human insulin[224, 240, 242–245]. The low immunogenicity of human insulin was confirmed in an uncontrolled study: only very low IgG levels were measured in new insulin patients treated with human insulin[247].

The situation is different when insulin-dependent diabetic patients are transferred to human insulin. Most of these patients already have high or low levels of insulin antibodies. The trend was for these levels to decrease after the patients had been treated for long enough periods with human insulin[193, 248, 249] or to remain constant[214, 220, 230]. The anti-insulin IgG levels were usually, however, too low to affect treatment – e. g. to necessitate changing the dose. Similar findings were made for IgE type antibodies[240, 242].

When manufacturing insulin by genetic engineering, there is in theory the possibility of contamination with peptides from E. coli, the organism used to produce the insulin. Serum samples from patients treated with this type of insulin were therefore investigated for antibodies to E. coli peptides. None were found[250a]. This was, however, a surprising finding, since 85% of all persons have suffered an E. coli infection at some time and show an immediate reaction when tested intracutaneously with E. coli peptides[143, 250].

β) Antigenicity

In addition to the concentration of circulating antibodies, the affinity of these antibodies to insulin also determines the total amount of insulin bound. The antibodies are composed of a variety of different molecular structures, the specificity of which may be directed at different areas of the insulin molecule. It is therefore clear that although the antibodies to one species during treatment are specific to the molecule involved, they also have a degree of crossreactivity with insulin molecules of other species. The intensity of the cross-reaction usually varies from individual to individual. If it is assumed that antibodies are formed to the „specifically human" areas of the human insulin molecule, A-8–A-10 and B-30, less binding of human insulin to antibodies formed against animal insulin can be expected. In the light of the above, however, it will be no surprise that consistent findings were not made. For example, there were groups of patients treated with porcine and bovine insulin with insulin allergy, who reacted differently to intracutaneous testing with different insulins. Some reacted to porcine, bovine and human insulin, the reaction to human insulin being weaker than to bovine insulin in most cases, but similar to or weaker than the reaction to porcine insulin[251]. Some patients reacted only to bovine insulin, some only to bovine and porcine insulin[131, 242, 243, 250, 252–255].

The binding of bovine, porcine and human insulin to IgE antibodies in allergic patients decreased progressively in this order[256]. Serum binding to IgG with human insulin was similar to that of porcine insulin[112, 113, 126, 143, 232, 241–243, 250, 255, 257] and usually lower than that of bovine insulin[213]. Only in rare cases was there no binding of bovine insulin, but binding of porcine and human insulin[250]. Binding was often lower with human insulin than with bovine and porcine insulin in patients with high antibody levels (IgG) coupled with resistance[254, 258].

The leucocyte migration inhibition test revealed no inhibition by various insulins[250, 252]. High affinity and low affinity binding constants (IgG) decreased in patients changed from bovine to human insulin[248]. The binding of insulins from different species to IgE antibodies was identical in patients previously treated with only human insulin[242]. Radioimmunoassay revealed no differences between various human insulins[115] and porcine insulins[124].

Human insulin therefore tends to bind less to IgG and IgE, which suggests that it would be of therapeutic advantage to change problem cases on animal insulins to human insulin.

γ) General Tolerance

This section covers some of the ground dealt with in the two previous chapters, since externally visible intolerance to insulin preparations is often connected with immunological processes.

Most of the reports on side effects include only small numbers of patients, since most of the cases under review were actually receiving the insulin preparations as treatment, and not within the framework of a clinical study.

As may be expected from previous sections, a uniform picture also did not emerge here.

On the one hand some patients who were allergic to animal insulins were free from symptoms after being changed to human insulin[256, 259–263] or at least the symptoms improved[119, 164]. Densensitisation with human insulin was possible, even in cases where a generalised allergy to human insulin was present[241, 265, 266a, 266b].

Skin tests are not necessarily an absolute indicator of the response of patients to particular insulins. It was possible to give successful subcutaneous treatment, even if the results of the intracutaneous testing – which is very sensitive – were positive[247]. One interesting case was reported in which treatment with NPH insulin was tolerated without problems, even though the patient had a positive reaction to regular human insulin in the skin test[269].

There were, however, also patients in whom skin testing with or changing treatment to human insulin was unsuccessful, producing the same results as with animal insulins[252, 256, 260, 263, 266–268].

It is no surprise that there are patients who initially tolerate human insulin well, but show a positive reaction in the skin test after some time[225]. Human insulin was also given to patients with antibody-induced resistance to animal insulins. Dose reductions were possible in these cases after transferring the patients to human insulin[119, 132, 270]. A very rare case with insulin resistance, even to human insulin, in the absence of high insulin-binding antibody levels, was managed with intravenous insulin and des-phe insulin[205].

Where antibodies to insulin receptors were present, changing to subcutaneous injections of human insulin brought no improvement, as expected[271]. The blood sugar levels were, however, able to be brought down with intravenous insulin[119].

3. Indications and Outlook for the Future

In accordance with clinical findings made so far, the principal indication for human insulin is intolerance to animal insulins[272–275]. Where efficacy and tolerance to animal insulins are satisfactory, substitution with human insulin appears to be inappropriate[276].

New insulin patients, however, responded well to human insulin[234, 272, 275] and it is recommended to treat those patients with human insulin from the very beginning.

Human insulin will also be useful for short-term treatment when (re)stabilising patients[272].

Human insulin will not revolutionise the treatment of insulin-dependent diabetes mellitus sufferers, but it is one further step along the road to better and more tolerable methods of treatment[168, 183, 277, 278, 279].

C. References

1. Smith, L. F.: Diabetes *21* (Suppl 2), 457–460 (1972)
2. Chan, S. J., Keim, P., Steiner, D. F.: Proc. Natl. Acad. Sci. USA *73,* 1964–1968 (1976)
3. Shield, D., Blobel, G.: Proc. Natl. Acad. Sci., USA *74,* 2059–2063 (1977)
4. Ullrich, A., Shine, J., Chrigwin, J., Pictet, R., Tischer, E., Rutter, W. J., Goodman, H. M.: Science *196,* 1313–1319 (1977)
5. Lomedico, P. T., Chan, S. J., Steiner, D. F., Saunders, G. F.: J. Biol. Chem. *252,* 7971–7978 (1977)
6. Cordell, B., Bell, G., Tischer, E., De Noto, F. M., Ullrich, A. Pictet, R., Rutter, W. J., Goodman, H. M.: Cell *18,* 533 (1979)
7. Lomedico, P., Rosenthal, N., Efstratiadis, A., Gilbert, W., Kolodner, R., Tizard, R.: Cell *18,* 545 (1979)
8. Chan, S. J., Noyes, B. E., Agarwal, K. L., Steiner, D. F.: Proc. Natl. Acad. Sci., USA *76,* 5036–5040 (1979)
9. Sures, I., Goeddel, D., Gray, A., Ullrich, A.: Science *208,* 57 (1980)
10. Nicol, D. S. H. W., Smith, L. F.: Nature *187,* 483–485 (1960)
11. Oyer, P. E., Cho, S., Peterson, J. D., Steiner, D. F.: J. Biol. Chem. *246,* 1375 (1971)
12. Ko, A. S. C., Smyth, D. G., Markussen, J., Sundby, F.: Eur. J. Biochem. *20,* 190 (1971)
13. Milstein, C., Brownlee, C. G., Harrison, T. M., Mathews, M. B.: Nature (London) New Biol. *239,* 117 (1972)
14. Blobel, G., Dobberstein, B.: J. Cell. Biol. *67,* 835 (1975)
15. Emdin, S. O., Dodson, G. G., Cutfield, J. M., Cutfield, S. M.: Diabetologia *19,* 174–182 (1980)
16. Howell, S. L.: The Insulin Storage Granule; p. 155, in: The Secretory Granule, (eds.) Poisner, A. M., Trifaro, J. M.: Elsevier Medical 1982
17. Chance, R. E.: Diabetes *21* (Suppl 2) 461–467 (1972)
18. Chance, R. E., Ellis, R. M., Bromer, W. W.: Science *161,* 165–167 (1968)
19. Naithani, V. K., Steffens, G. J., Tager, H. S., Buse, G., Rubenstein, A. H., Steiner, D. F.: Hoppe Seyler's Z. Physiol. Chem. *365,* 571–575 (1984)
20. Peterson, J. D., Nehrlich, S., Oyer, P. E., Steiner, D. F.: Biol. Chem. *247,* 4866–4871 (1972)
21. Dodson, E. J., Dodson, G. G., Reynolds, C. D., Vallely, D.: p. 9, in: Insulin, (eds.) Brandenburg, D., Wollmer, A., W. de Gruyter, Berlin–New York 1980
22. Chawdhury, S., Dodson, G., Reynolds, C., Tolley, S.: Diabetologia *22,* 384 (1982)
23. Tager, H. S., Given, B., Baldwin, D., Mako, M., Rubenstein, A., Olefsky, J., Kobayashi, M., Kollermann, O., Poucher, P.: Nature *281,* 122–125 (1979)
24. Shoelson, S., Haneda, M., Blix, P., Nanjo, A., Sanke, T., Inouye, K., Steiner, D. F., Rubenstein, A. H., Tager, H. S.: Nature *302,* 540–543 (1983)
25. Shoelson, S., Fickova, M., Haneda, M., Nahum, A., Musso, G., Kaiser, E. T., Rubenstein, A. H., Tager, H. S.: Proc. Natl. Acad. Sci. USA *80,* 7390–7394 (1983)
26. Tager, H. S.: Diabetes *33,* 693–699 (1984)
27. Kwok, S. C. M., Steiner, D. F., Rubenstein, A. H., Tager, H. S.: Diabetes *32,* 872–875 (1983)
28. Haneda, M., Chan, S. J., Kwok, S. C. M., Rubenstein, A. H., Steiner, D. F.: Proc. Natl. Acad. Sci. USA *80,* 6366–6370 (1983)
29. Kimmel, J. R., Pollock, H. G.: Diabetes *16,* 687–694 (1967)

30. Inouye, K., Watanabe, K., Tochino, Y., Kobayashi, M., Shigeta, Y.: Biopolymers *20,* 1845–1858 (1981)
31. Tager, H., Thomas, N., Assoian, R., Rubenstein, A., Sackow, M., Olefsky, J., Kaiser, E. T.: Proc. Natl. Acad. Sci. USA *77,* 3181–3185 (1980)
32. Wollmer, A., Strassburger, W., Glatter, V., Dodson, G. G., Mc Call, M., Danho, W., Brandenburg, D., Gattner, H.-G., Rittel, W.: Hoppe Seyler's Z. Physiol. Chem. *362,* 581–592 (1981)
33. Assoian, R. K., Thomas, N. L., Kaiser, E. T., Tager, H. S.: Proc. Natl. Acad. Sci. *79,* 5147–5151 (1982)
34. Robbins, D. C., Blix, P. M., Rubenstein, A. H., Kanazawa, Y., Kosaka, K., Tager, H. S.: Nature *291,* 679–681 (1981)
35. Robbins, D., Shoelson, S., Rubenstein, A., Tager, H. S.: J. Clin. Invest. *73,* 714–719 (1984)
36. Steiner, D. F., Quinn, P. S., Chan, S. J., Marsh, J., Tager, H. S.: Ann. NY Acad. Sci., *343,* 133–147 (1980)
37. Klostermeyer, H., Zahn, H.: Handbuch der experim. Pharmazie Bd. XXXII/1 ed. E. Dörzbach, Springer Verlag 1971
38. Geiger, R.: Chemiker Zeitung *100,* 111–129 (1976)
39. see[126]
40. USP XXI (1985) United States Pharmacopoeia
41. Mirsky, I. A., Jinks, R., Perisutti, G.: J. clin. Invest. *42,* 1869 (1963)
42. Smith, L. F., Biochim. Biophys. Acta *82,* 231 (1964)
43. Jackson, R. L., Shney, E. W., Grinnan, E. L., Ellis, R. M.: Diabetes *18,* 206 (1969)
44. Shapcott, D., O'Brien, D.: Diabetes *19,* 831 (1970)
45. Dixon, G. H., Wardlaw, A. C.: Nature *188,* 721 (1960)
46. Du, Y.-C., Zhang, Y.-S., Lu, Z.-X., Tsou, C.-L.: Sci. Sin. *10,* 84 (1961)
47. Katsoyannis, P. G., Tometsko, A., Fukuda, K.: J. Amer. Chem. Soc. *85,* 2863 (1963)
48. Dixon, G. H.: Exerpta. Med. Found. Int. Congr. Ser. *83,* 1207 (1964)
49. Meienhofer, J., Schnabel, E., Bremer, H., Brinkhoff, O., Zabel, R., Sroka, W., Klostermeyer, H., Brandenburg, D., Okuda, T., Zahn, H.: Z. Naturforsch. *186,* 1120 (1963)
50. Katsoyannis, P. G., Fukuda, K., Tometsko, A., Suzuki, K., Tilak, M.: J. Amer. Chem. Soc. *86,* 930 (1964)
51. Kung, Y.-T., et al., Sci. Sin. *14,* 1710 (1965)
52. Katsoyannis, P. G., Trakatellis, A. C., Zalut, C., Johnson, S., Tometsko, A. Schwartz, G., Ginos, J.: Biochemistry *6,* 2656–2668 (1967)
53. Sieber, P., Kamber, B., Hartmann, A., Jöhl, A., Riniker, B., Rittel, W.: Helv. Chim. Acta *57,* 2617 (1974)
54. Kamber, B.: Helv. Chim. Acta *56,* 1370 (1973)
55. Sieber, P., Iselin, B.: Helv. Chim. Acta *51,* 622 (1968)
56. Büllesbach, E. E., Brandenburg, D.: Synthesis of insulin analogues. In: Basic and clinical aspects of immunity to insulin; ed. Keck, K. and Erb, P.; p. 395–419, W. de Gruyter, Berlin–New York, 1981
57. Brandenburg, D., Saunders, D.: Pancreatic hormones. In: Amino acids, Peptides and Proteins (Sheppard, R. C., Senior Reporter), Royal Soc. Chemistry, London *12,* 496–513 (1981)
58. Märki, F., De Gasparo, M., Eisler, K.: Hoppe Seyler's Z. Physiol. Chem. *11,* 1619–1632 (1979)
59. Steiner, D. F., Clark, J. L.: Proc. Natl. Acad. Sci. USA *60,* 622 (1968)
60. Yanaihara, N., Sakagami, M., Yanaihara C.: see *21;* p. 81–89
61. Kemmler, W., Peterson, J. D., Steiner, D. F.: J. Biol. Chem. *246,* 6786–6791 (1971)
62. Lindsay, D. G.: FEBS Lett. *21,* 105 (1971)
63. Brandenburg, D., Wollmer, A.: Hoppe Seyler's Z. Physiol. Chem. *354,* 613 (1973)
64. Geiger, R., Obermeier, R.: Biochem. Biophys. Res. Commun. *55,* 60 (1973)
65. Brandenburg, D., Schermutzki, W., Zahn, H.: Hoppe Seyler's Z. Physiol. Chem. *354,* 1521 (1973)
66. Ruttenberg, M. A.: Science *177,* 623–626 (1972)
67. The Shanghai Insulin Research Group: Sci. Sin. *16,* 61–70 (1973)
68. Weitzel, G., Bauer, F.-U., Eisele, K.: Hoppe Seyler's Z. Physiol. Chem. *357,* 187–200 (1976)
69. Obermeier, R., Geiger, R.: Hoppe Seyler's Z. Physiol. Chem. *357,* 759–767 (1976)

70. Sondheimer, E., Holley, R.: J. Amer. Chem. Soc. *76,* 2467 (1954)
71. König, W., Volk, A.: Chem. Ber. *110,* 1–11 (1977)
72. König, W., Geiger, R.: Chem. Ber. *103,* 788 (1970)
73. Fruton, J.: Proteinase-Catalyzed Synthesis of Peptide Bonds, in Advances in Enzymology *53,* 239–306, Ed. A. Meister, Wiley & Sons, 1982.
74. Jakubke, H. D., Kuhl, P.: Die Pharmazie *37,* 89–106 (1982)
75. Inouye, K., Watanabe, K., Morihara, K., Tochino, Y., Kanaya, T., Emura, J., Sakakibara, S.: J. Amer. Chem. Soc. *101,* 751–752 (1979)
76. Morihara, K., Oka, T., Tsuzuki, H.: Nature, *280,* 412–413 (1979)
77. Gattner, H.-G., Danho, W., Naithani, V. K.: in: (21) p. 117–123
78. Morihara, K., Oka, T., Tochino, H., Kanaya, T.: Biochem. Biophys. Res. Commun. *92,* 396–402 (1980)
79. Obermeier, R., Seipke, G., Grau, U.: Human Insulin Semisynthesis, in: Neue Insuline, Proc. of the Int. Symposium, Freiburg i. Br., ed. Kerp, L., Schlüter, K. G., p. 33–37, 1982
80. Markussen, J.: ibid. 38–44
81. Jonczyk, A., Gattner, H.-G.: Hoppe Seyler's Z. Physiol. Chem. *362,* 1591 (1981)
82. Rose, K., De Pury, H., Offord, R. E.: Biochem. J. *211,* 671–676 (1983)
83. Schmitt, E., Gattner, H.-G.: Hoppe Seyler's Z. Physiol. Chem. *359,* 799 (1978)
84. Masaki, T., Nakamura, K., Isono, M., Soejima, M.: Agric. Biol. Chem. *42,* 1443 (1978)
85. Obermeier, R., Seipke, G.: Process Biochemistry *19,* 29–32 (1984)
86. Bloom, S. R., Adrian, T. E., Barnes, A. J., Polak, J. M.: Lancet, i, 14 (1979)
87. Kobayashi, M., Okgaku, S., Iwasaki, M., Shigata, G., Oka, T., Morihara, K.: Diabetes *30,* 519 (1981)
88. Neubauer, H. P., Obermeier, R., Schnorr, G.: Diabetologia *27,* 129–131 (1984)
89. Kerp, L., Steinhilber, S., Kasemir, H., Han, J., Henrichs, H. R., Geiger, R.: Diabetes *23,* 651–656 (1974)
90. Rose, K., Gladstone, J., Offord, R. E.: Biochem. J. *220,* 189–196 (1984)
91. Markussen, J., Schaumburg, K.: Diabetologia *23,* 185 (1982)
92. Burnett, J. P., Commercial Production of Recombinant DNA-Derived Products, in: Experimental Manipulation of Gene Expression, ed. M. Inouye, Academic Press, p. 259–277 (1983)
93. Chance, R. E., Kroeff, E. P., Hoffmann, J. A.: in: ed. J. L. Gueriguian, Insulins, Growth Hormones and Recombinant DNA Technology, Raven Press, New York, p. 71 (1981)
94. Wengenmayer, F.: Angew. Chem. *95,* 874–891 (1983)
95. Goeddel, D. V. Kleid, D. G., Bolivar, F., Heyneken, H. L., Yansura, D. G., Crea, R., Hirose, T., Kraszewski, A., Itakura, K., Riggs, A. D.: Proc. Natl. Acad. Sci., USA *76,* 106–110 (1979).
96. Crea, R., Kraszewski, T., Hirose, K., Itakura, K.: Proc. Nat. Acad. Sci., USA *75,* 5765 (1978)
97. Gattner, H. G., Krail, G., Danho, W., Knorr, R., Wieneke, H.-J., Büllesbach, E. E., Schartmann, B., Brandenburg, D., Zahn, H.: Hoppe Seyler's Z. Physiol. Chem. *362,* 1043–1049 (1981)
98. Schartmann, B., Gattner, H. G., Danho, W., Zahn, H.: Hoppe Seyler's Z. Physiol. Chem. *364,* 179–186 (1983)
99. Swamy, K. H. S., Goldberg, F.: J. Bacteriol. *149,* 1027–1033 (1982)
100. Burgermeister, W., Enzmann, F., Schöne, H.-H.: The Isolation of Insulin form the Pancreas, in: ed. G. V. R. Born, O. Eichler, A. Farah, H. Herken, A. D. Welch, Vol. XXXII/2, Ed. H. Hasselblatt, F. v. Bruchhausen, Handbuch der experim. Pharmakologie, Neue Serie, p. 715–727 (1975). Springer-Verlag, Berlin–Heidelberg–New York
101. McLeod, A., Wood, S. P.: J. Chromatography *285*, 319–331 (1984)
102. Damgaard, U., Markussen, J.: Horm. Metab. Res. *11,* 580–581 (1979)
103. Chance, R. E., Hoffmann, J. A., Kroeff, E. P., Johnson, M. G., Schirmer, E. W., Bromer, W. W., Ross, M. J., Wetzel, R., in: Proc. 7th American Peptide Symposium, ed. D. H. Rich, E. Gross, Pierce Chemical Company, Rockford, Ill., p. 721–728 (1981)
104. Grau, U.: Diabetes, 1985, submitted
105. Baker, R. S., Ross, J. W., Schmidtke, J. R., W. C. Smith: Lancet, 1139–1142 (1981)
106. Cooper, J. F., Levin, J., Wagner, H.: J. Lab. Clin. Med. *78,* 138–148 (1971)
107. Chawdhury, S. A., Dodson, E. J., Dodson, G. G., Reynolds, C. D., Tolley, S. P., Blundell, T. L., Cleasby, A., Pitts, J. E., Tickle, I. J., Wood, S. P.: Diabetologia *25,* 460 (1983)

108. Lotz, N., Lacher, F., Bachmann, W.: In-vitro Bindungsstudien von biosynthetischem Humaninsulin an Erythrozyten, in: Neue Insuline, (ed.) Petersen et al., p. 105, Freiburg, Freiburger graph. Betriebe (1982)
109. Schlüter, K. J., Petersen, K.G., Burmeister, P., Kerp, L.: Unterschiedliche Rezeptorbindung verschiedener Tracer-Insuline, in: Neue Insuline, (ed.) Petersen et al., p. 118, Freiburg, Freiburger graph. Betriebe (1982)
110. Keefer, L. M., Piron, M.-A., De Meyts, P.: Diabetes Care *4,* 209 (1981)
111. Bachmann, W., Sieger, C., Lacher, F., Lotz, N.: Diabetes Care *4,* 215 (1981)
112. Schernthaner, G., Borkensteiner, M., Mathä, R., Mayr, W. R., Hohenecker, J., Prager, R., Schober, E., Susani, M.: Semisynthetisches und biosynthetisches Humaninsulin: Immunogenitätsuntersuchungen nach Langzeittherapie bei Typ-I Diabetikern und Bindungsanalysen an präformierten IgG-Insulinantikörpern und Insulinrezeptoren, in: Neue Insuline, (ed.) Petersen et al., p. 140, Freiburger graph. Betriebe (1982)
113. Petersen, K.-G., Schlüter, K., Steinhilber, S., Kerp, L.: Diabetes Care *4,* 248 (1981)
114. Prager, R., Schernthaner, G.: Diabetes Care *5* (Suppl. 2), 104 (1982)
115. Chance, R., Kroeff, E. P., Hiffmann, J. A., Frank, B. H.: Diabetes Care *4,* 147 (1981)
116. Carpentier, J. L., Halban, Ph. A., Renold, A. E., Orci, L.: Diabetes Care *4,* 220 (1981)
117. Fußgänger, R. D., Ditschuneit, H. H., Martini, H., Wiebauer-de Lenardis, H., Etzrodt, H., Thun, Ch., Ditschuneit, H., Pfeiffer, E. F., Enzmann, F.: Diabetes Care *4,* 228 (1981)
118. Olefsky, J.: Diabetes Care *4,* 244 (1981)
119. Schlüter, K. J., Kerp, L.: Diabetes Care *5* (Suppl. 2), 152 (1982)
120. Zick, R., Hürter, P., Meyer, B., v. Schütz, W., Holle, W., Dwenger, A., Mitzkat, H. J.: Münch. med. Wschr. (Suppl. 1) *125,* 97 (1983)
121. Prager, R., Schernthaner, G.: Diabetologia *25,* 235 (1983)
122. Leiper, J. M.: Diabetes Res. *1,* 27 (1984)
123. Sonne, D., Foley, J. E., Gliemann, J.: Diabetes Care *4,* 250 (1981)
124. Wieringa, T., van Putten, J. P. M., Van Dijk-Besling, M., Krans, H. M.: Diabetes Care *4,* 254 (1981)
125. Khokher, M. A., Dandona, P.: Diabetes Care *5* (Suppl. 2), 102 (1982)
126. Märki, F., Albrecht, W.: Diabetologia *13,* 293 (1977)
127. Zapf, J., Froesch, E. R.: Diabetes Care *4,* 257 (1981)
128. Fehlmann, M., Le Marchand-Brustel, Y., Dolais-Kitabgi, J., Morni, O., Freychet, P.: Diabetes Care *4,* 223 (1981)
129. Gammeltoft, S.: Diabetes Care *4,* 235 (1981)
130. Verspohl, E. J., Ammon, H. P. T.: Diabetes Care *4,* 252 (1981)
131. Waldhäusl, W., Kastner, G., Komjati, M., Bratusch-Marrain, P.: Diabetes Care *4,* 205 (1981)
132. Schlüter, K. J., Petersen, K.-G., Blachnitzky, E. O., Steinhilber, S., Kerp, L.: Abstr. Dt. Diab. Kongreß, Paper 65 (1979)
133. Wilson, D. E.: Ann. Int. Med., 98, 219 (1983)
134. Halban, P., Berger, M., Gjinovci, A., Renold, A. E.: Diabetes Care *4,* 238 (1981)
135. Gamst-Andersen, H., Hjortkjaer, R. K., Lorenzen, F. H., Ashby, R.: Act. pharm. et tox. *52,* 261 (1983)
136. Damm-Jorgensen, K., Hafstein Wolffbrandt, K., Weis, J. U.: Act. pharm. et tox. *52,* 268 (1983)
137. Ebihara, A., Kondo, K., Ohashi, K., Kosaka, K., Kuzuya, T., Matsuda, A.: Diabetes Care *6* (Suppl. 1), 17 (1983)
138. Ebihara, A., Kondo, K., Ohashi, K., Kosaka, K., Kuzuya, T., Tarni, S.: Diabetes Care *5* (Suppl. 2), 35 (1982)
139. Owens, D. R., Jones, M. K., Birtwell, A. J., Jones, I. R., Hayes, T. M., Heding, L. G.: Diabetes Care *6* (Suppl. 1), 13 (1983)
140. Owens, D. R., Jones, M. K., Hayes, T. M., Heding, L. G., Alberti, K. G. M. M., Home, P. D., Burrin, J. M., Newcombe, R. G.: Brit. Med. J. *282,* 1264 (1981)
141. Owens, D. R., Jones, M. K., Hayes, T. M., Heding, L. G., Alberti, K. G. M. M., Home, P. D.: Lancet, 118 (1981)
142. Laube, H., Velcovsky, H. G., Mäser, E., Federlin, K.: Vergleichende Untersuchung zur Wirkung von unterschiedlichen Humaninsulinen bei gesunden Probanden, in: Neue Insuline, (ed.) Petersen et al., p. 61, Freiburg, Freiburger graph. Betriebe (1982)

143. Federlin, K., Laube, H., Velcovsky, H. G.: Diabetes Care *4,* 170 (1981)
144. Galloway, J. A., Root, M. A., Berstrom, R., Spradlin, C. T., Howey, D. C., Fineberg, S. E., Jackson, R. L.: Diabetes Care *5* (Suppl. 2), 13 (1982)
145. Raptis, S., Hadjidakis, D., Enzmann, F., Raptis, A., Diamantopoulos, E.: Diabetes Care *5* (Suppl. 2), 93 (1982)
146. Gyaram, H., Wahl, K., Ermler, R., Lebender, A., Bottermann, P.: Pharmakokinetik und Wirkprofile von Des-Phe-Insulinen sowie biosynthetischen und semisynthetischen Humaninsulinen, in: Neue Insuline, (ed.) Petersen et al., p. 187, Freiburg, Freiburger graph. Betriebe (1982)
147. Bottermann, P., Gyaram, H., Wahl, K., Ermler, R., Lebender, A.: Diabetes Care *4,* 168 (1981)
148. Sonnenberg, G. E., Kemmer, F. W., Cüppers, H.-J., Berger, M.: Diabetes Care *6* (Suppl. 1), 35 (1983)
149. Kemmer, F. W., Sonnenberg, G., Cüppers, H.-J., Berger, M.: Diabetes Care *5* (Suppl. 2), 23 (1982)
150. Pickup, J. C., Bilous, R. W., Viberti, G. G., Keen, H., Jarrett, R. J., Glynne, A., Cauldwell, J., Root, M., Rubenstein, A. H.: Diabetes Care *5* (Suppl. 2), 29 (1982)
151. Owens, D. R., Jones, I. R., Birtwell, A. J., Burge, C. T. R., Luzio, S.: Diabetologia *26,* 261 (1984)
152. Bottermann, P., Gyaram, H., Wahl, K., Ermler, R., Lebender, A.: Diabetes Care *5* (Suppl. 2), 43 (1982)
153. Weinges, K., Ehrhardt, M., Nell, G., Enzmann, F.: Diabetes Care *5* (Suppl. 2), 67 (1982)
154. Raptis, S., Karaiskos, C., Enzmann, F., Hatzidakis, D., Zoupas, Ch., Souvatzoglou, A., Diamantopoulos, E., Moulopoulos, S.: Diabetes Care *4,* 155 (1981)
155. Schlüter, K. J., Petersen, K.-G., Sontheimer, J., Enzmann, F., Kerp, L.: Diabetes Care *5* (Suppl. 2), 78 (1982)
156. Schlüter, K. J., Petersen, K.-G., Burmeister, P., Kerp, L.: Unterschiedliche Wirkung von Human- und Schweineinsulin, in: Neue Insuline, (ed.) Petersen et al., p. 86, Freiburg, Freiburger graph. Betriebe (1982)
157. Schlüter, K. J., Petersen, K.-G., Borsche, A., Hobitz, H., Kerp, L.: Horm. Metabol. Res. *14,* 109 (1982)
158. Schlüter, K. J., Enzmann, F., Kerp, L.: Horm. Metabol. Res. *15,* 271 (1983)
159. Arias, P., Kerner, W., Navascués, I., Schäfauer, G., Pfeiffer, E. F.: Klin. Wschr. *62,* 1145 (1984)
160. Skyler, J. S.: Diabetes Care *5* (Suppl. 2), 181 (1982)
161. Frost, D. P., Srivastava, M. C., Jones, R. H., Nabarro, J. D. N., Sönksen, P. H.: Postgrad. Med. J. *49,* 949 (1973)
162. Home, P. D., Shepherd, G. A. A., Noy, G., Massi-Benedetti, M., Hanning, I., Burrin, J. M., Alberti, K. G. M. M.: Diabetes Care *6* (Suppl. 1), 23 (1983)
163. Massi-Benedetti, M., Burrin, J. M., Capaldo, B., Alberti, K. G. M. M.: Diabetes Care *4,* 163 (1981)
164. Home, P. D., Massi-Benedetti, M., Shepherd, G. A. A., Hanning, I., Alberti, K. G. M. M., Owens, D. R.: Diabetologia *22,* 41 (1982)
165. Müller, R., Berger, W., Wick, H., Keller, U.: Horm. Metabol. Res. *16,* 271 (1984)
166. Gerlis, L. S., Adeniyi-Jones, R., Jones, R. H., Sönksen, P. H., Barnes, G. D.: Die Auswirkung von Änderungen des Blutzuckers auf den Metabolismus und die Verteilung von Humaninsulin beim Menschen, in: Neue Insuline, (ed.) Petersen et al., p. 66, Freiburg, Freiburger graph. Betriebe (1982)
167. Petersen, K.-G., Kerp, L.: Personal Communication, data on file Hoechst AG
168. Galloway, J. A.: Marsden, J. H., Root, M. A., Spradlin, C. T.: A Review of Clinical Trials of Biosynthetic Human Insulin in the United States, in: Neue Insuline, (ed.) Petersen et al., p. 175, Freiburg, Freiburger graph. Betriebe (1982)
169. Galloway, J. A., Spradlin, C. T., Root, M. A., Fineberg, S. E.: Diabetes Care *4,* 183 (1981)
170. Viberti, G. C., Pickup, J. C., Keen, H., Bilous, R. W., Mattock, M., Jarrett, R. J., Glynne, A., Rogers, R.: Diabetes Care *4,* 175 (1981)
171. Keen, H., Glynne, A., Pickup, J. C., Vilberti, G. C., Bilous, R. W., Jarrett; R. J., Marsden, R.: Lancet *129,* 398 (1980)

172. Weinges, K., Ehrhardt, M., Enzmann, F.: Diabetes Care *4,* 180 (1981)
173. Rosak, C., Schöffling, K.: Personal Communication, data on file Hoechst AG
174. Petersen, K.-G., Schlüter, K. J., Burmeister, P., Wehrli, A., Kerp, L.: Die Wirkung von Human- und Schweineinsulin auf die Katecholaminausschüttung in der Hypoglykämie, in: Neue Insuline, (ed.) Petersen et al., p. 73, Freiburg, Freiburger graph. Betriebe (1982)
175. Rosak, C., Althoff, P.-H., Schöffling, K.: Untersuchungen zur blutzuckersenkenden, antilipolytischen, antiketogenen, Kalium- und Phosphat-senkenden Eigenschaft von Humaninsulin, Schweineinsulin und Des-Phe-Insulin, in: Neue Insuline, (ed.) Petersen et al., p. 79, Freiburg, Freiburger graph. Betriebe (1982)
176. Rosak, C., Althoff, P.-H., Enzmann, F., Schöffling, K.: Diabetes Care *5* (Suppl. 2), 82 (1982)
177. Drost, H., et al.: Untersuchungen zur hormonellen Regulation von Hypoglykämien induziert durch biosynthetisches Humaninsulin, in: Neue Insuline, (ed.) Petersen et al., p. 100, Freiburg, Freiburger graph. Betriebe (1982)
178. Petersen, K.-G., Schlüter, K. J., Kerp, L.: Diabetes Care *5* (Suppl. 2), 90 (1982)
179. Polonsky, K., Bergenstal, R., Pons, G., Schneider, M., Jaspan, J., Rubenstein, A: New. Engl. J. Med. *307,* 1106 (1982)
180. Bieger, W. P.: Therapiewoche *34,* 1665 (1984)
181. Chisholm, D., Kraegen, E. W., Hewett, M. J., Lazarus, L.: Horm. Metabol. Res. *15,* 415 (1983)
182. De Fronzo, R. A., Felig, P., Ferrannini, E., Wahren, J.: Am. J. Physiol. *238,* 421 (1980)
183. Etzwiler, D. D.: Diabetes Care *6* (Suppl. 1), 66 (1983)
184. Hoogen v. d., F. H. J., Hulst, S. G. Th., Zweens, J.: Diabetes Care *5* (Suppl. 2), 71 (1982)
185. Vick, R. L., Todd, E. P., Luedke, P. W.: J. Pharmacol. Exp. Ther. *181,* 139 (1972)
186. Waldhäusl, W., Vierhapper, H., Bratusch-Marrain, P.: Unterschiedliche Suppression der endogenen Insulininkretion durch biosynthetisches Humaninsulin beim Menschen, in: Neue Insuline, (ed.) Petersen et al., p. 93, Freiburg, Freiburger graph. Betriebe (1982)
187. Kuzuya, H., Blix, P. M., Horwitz, D. L., Steiner, D. E., Rubenstein, A.: Diabetes *26,* 22 (1977)
188. Colagiuri, S., Kotowicz, M. A., Steinbeck, A. W., Kidson, W.: Diabetes Care *6* (Suppl. 1), 49 (1983)
189. Scott, R., Smith, J.: Diabetes Metabol. 9, 95 (1983)
190. Castillo, M., Nemery, A., Verdin, E., Lefebvre, P. J., Luyckx, A. S.: Eur. J. Clin. Pharmacol. *25,* 767 (1983)
191. Hildebrandt, P., Birch, K., Sestoft, L., Vølund, A.: Acta Med. Scand. *215,* 69 (1984)
192. Massi-Benedetti, M., Bueti, A., Calabrese, G., Zega, G., Brunetti, P.: Untersuchungen über die Wirkung von biosynthetischem Human-Insulin an Hand der Glukoseclamptechnik bei insulinabhängigem Diabetes, in: Neue Insuline, (ed.) Petersen et al., p. 53, Freiburg, Freiburger graph. Betriebe (1982)
193. Charles, M. A. Szekeres, A., Staten, M., Worcester, B., Walsh, K. M.: Diabetes Care *6* (Suppl. 1), 29 (1983)
194. Howey, D. C., Fineberg, S. E., Nolen, P. A., Stone, M. I., Gibson, R. G., Fineberg, N. S., Galloway, J. A.: Diabetes Care *5* (Suppl. 2), 73 (1982)
195. Staten, M., Worcester, B., Szekeres, A., Waldeck, N., Ascher, M., Walsh, K. M., Rizza, R., Gerich, J., Charles, M. A.: Metabolism *33,* 132 (1984)
196. Petzoldt, R., Finck, H.-D., Klein, E.: Über die Wirkung von Human-Insulin und Des-Phe-Insulin bei 8 Typ I-Diabetikern, in: Neue Insuline, (ed.) Petersen et al., p. 196, Freiburg, Freiburger graph. Betriebe (1982)
197. Kerp, L., Steinhilber, S. Kasemir, H., Han, J., Henrichs, H. R., Geiger, R.: Diabetes, *23,* 651 (1974)
198. Massi-Benedetti, M., Buetti, A., Mannino, D., Bellomo, G., Antonella, M. A., Calabrese, G., Zega, G., Brunetti, P.: Diabetes Care *7,* 132 (1984)
199. Mirouze, J., Benghernaout, O., Pham, T. C., Richard, J. L., Bringer, J.: Diabetes Care *6* (Suppl. 1), 40 (1983)
200. Beyer, J., Weber, Th., Schulz, G., Hassinger, W., Westerburg, A., Cordes, U.: Diabetes Care *4* 189 (1981)
201. Klier, M., Kerner, W., Torres, A. A., Pfeiffer, E. F.: Diabetes Care *4,* 193 (1981)

202. Szekeres, A., Worcester, B., Ascher, M. S., Tuxen, D., Heyendal, R., Walsh, K. M., Charles, M. A.: Diabetes Care *6,* 193 (1983)
203. Schusdziarra, M.: Münch. Med. Wschr. *125* (Suppl. 1), 35 (1983)
204. Weber, T., Beyer, J., Cordes, U., Schulz, G., Brandstetter, K., Diederich, Ch., Krause, U., Enzmann, F.: Erstbehandlung des juvenilen Diabetes mellitus mit biosynthetischem Humaninsulin, in: Neue Insuline, (ed.) Petersen et al., p. 210, Freiburg, Freiburger graph. Betriebe (1982)
205. Walter, H., Lotz, N., Mehnert, H., Bachmann, W.: Biosynthetisches Human-Insulin (BHI) in der Behandlung des Coma diabeticum, in: Neue Insuline, (ed.) Petersen et al., p. 216, Freiburg, Freiburger graph. Betriebe (1982)
205a. Vyver van de, F. L., Thenaers, G., De Leeuw, I.: Insulinresistenz behandelt mit kontinuierlicher intravenöser Infusion von semisynthetischem Humaninsulin, in: Neue Insuline, (ed.) Petersen et al., p. 222, Freiburg, Freiburger graph. Betriebe (1982)
206. Bachmann, W., Walter, H., Lotz, N., Mehnert, H.: Diabetes Care *5* (Suppl. 2), 161 (1982)
207. Bachmann, W., Lotz, N., Walter, H., Mehnert, H.: Münch. Med. Wschr. *125* (Suppl. 1), 80 (1983)
208. Lotz, N., Bachmann, W., Mehnert, H.: Diabetes Care *5* (Suppl. 2), 149 (1982)
209. Sonnenberg, G., Chantelau, E., Sundermann, S., Hauff, C., Berger, M.: Vergleich von semisynthetischem Human- und Schweineinsulin: Doppelblindstudie bei Typ-I Diabetikern mit der kontinuierlichen subkutanen Insulininfusion (CSII), in: Neue Insuline, (ed.) Petersen et al., p. 444, Freiburg, Freiburger graph. Betriebe (1982)
210. Renner, R., Hepp, K. D.: Intraperitoneale Humaninsulinzufuhr bei Typ I-Diabetes mit ungewöhnlichem Cortisolprofil, in: Neue Insuline, (ed.) Petersen et al., p. 232, Freiburg, Freiburger graph. Betriebe (1982)
211. Grau, U., Seipke, G., Obermeier, R., Thurow, H.: Stabile Insulinlösungen für automatische Dosiergeräte, in: Neue Insuline, (ed.) Petersen et al., p. 411, Freiburg, Freiburger graph. Betriebe (1982)
212. Mirouze, J., Monnier, L., Richard, J. L., Gancel, A., Soua, K. B.: Diabetes Care *5* (Suppl. 2), 60 (1982)
213. Holman, R. R., Steemson, J., Darling, P., Reeves, W. G., Turner, R. C.: Brit. Med. J. *288,* 665 (1984)
214. Lyngsøe, J., Vestermark, S.: Diabetes Care *6* (Suppl. 1), 53 (1983)
215. Karam, J., Brink, St., Clements, R., Miller, L., Raskin, Ph.: Diabetes Care *6* (Suppl. 1), 56 (1983)
216. Landgraf, R., Kammerer, S., Bock, T., Smolka, B., Ladik, T., Enzmann, F., Landgraf-Leurs, M. M. C.: Diabetes Care *5* (Suppl. 2), 39 (1982)
217. Renner, R., Vocke, K., Hepp, K. D.: Diabetes Care *5* (Suppl. 2), 53 (1982)
218. Sailer, D., Ludwig, Th., Kolb, S.: Diabetes Care *5* (Suppl. 2), 57 (1982)
219. Marre, M., Tabbi-Anneni, A., Tabbi-Anneni, H., Assan, R.: Diabetes Care *5* (Suppl. 2), 63 (1982)
220. Clark, A. J., Wiles, Ph. G., Leiper, J. M., Knight, G., Adeniyi-Jones, R. O., Watkins, P. J., Ward, J. D., MacCuish, A. C., Keen, H., Jones, R. H.: Diabetes Care *5* (Suppl. 2), 129 (1982)
221. Galloway, J. A., Peck, F. B., Fineberg, S. E., Spradlin, C. Th., Marsden, J. H., Allemenos, D., Ingulli-Fattic, J.: Diabetes Care *5* (Suppl. 2), 135 (1982)
222. Beyer, J., Enzmann, F., Lauerbach, M., Althoff, P., Bachmann, W., Brandstetter, K., Cordes, U., Dannehl, A., Düntsch, G., Haas, R., Heun, K., Kerp, L., Lotz, N., Mehnert, H., Mitzkat, H.-J., Petersen, K., Rosak, C., Sailer, D., Schlüter, K., Schöffling, K., Schröder, K. E., Schumm, P., Weber, T.: Diabetes Care *5* (Suppl. 2), 140 (1982)
223. Beyer, J., Haas, R., Enzmann, F., Lauerbach, M., Radenacker, J., Althoff, P.-H., Bachmann, W., Brandstetter, K., Cordes, U., Dannehl, A., Düntsch, G., Heun, K., Kerp, L., Lotz, N., Mehnert, H., Mitzkat, H. J., Petersen, K., Rosak, C., Sailer, D., Schlüter, K., Schöffling, K., Schröder, K. E., Schumm, P.: Münch. Med. Wschr. *125* (Suppl. 1), 69 (1983)
224. Javicoli, M., Di Mario, U., Coronel, G. A., Dawud, A. M., Ardnini, P., Leonardi, M.: Diabetes Care *7,* 128 (1984)
225. Schlüter, K. J., Petersen, K.-G., Kerp, L.: Deutsch. Med. Wschr. *107,* 10 (1982)
226. Etti, H., Haupt, E.: Münch. Med. Wschr. *125* (Suppl. 1), 53 (1983)

227. Landgraf, R., Landgraf-Leurs, M. M. C.: Münch. Med. Wschr. *125* (Suppl. 1), 63 (1983)
228. Schlüter, K. J., Kerp, L., Enzmann, F., Rademacher, J., Lauerbach, M., Bachmann, W., Beyer, J., Bierich, J. R., Federlin, K., Gries, F. A., Haas, R., Heinze, E., Hübinger, A., Mäser, E., Mehnert, H., Nothjunge, J., Steinhilber, S., Thon, A., Velcovsky, H. G.: Münch. Med. Wschr. *125* (Suppl. 1), 108 (1983)
229. Egstrup, K., Olsen, J.: Lancet, 222 (1982)
230. Peacock, I., Tattersall, R. B., Taylor, A., Douglas, C. A., Reeves, W. G.: Lancet, 149 (1983)
231. Mofitt, P. S., Colagiuri, St., Miller, J. J., Hall, Ch. A.: Med. J. Aust. *140,* 200 (1984)
232. Drury, R., Keenan, P., McEvoy, M., Cregan, D., Drury, M. I., Reeves, W. G.: Ir. J. Med. Sci. *152,* 430 (1983)
232a. Teuscher, A.: Schweiz. Med. Wschr. *109,* 743 (1979)
233. Mann, N. P., Johnston, D. I., Reeves, G., Murphy, M. A.: Brit. Med. J. *287,* 1580 (1983)
234. Greene, S. A., Smith, M. S., Cartwright, B., Baum, J. D.: Brit. Med. J. *287,* 1578 (1983)
235. Hürter, P., Stolzenbach, K., v. Schütz, W., Kubel, R., Zick, R.: Münch. Med. Wschr. *125* (Suppl. 1), 91 (1983)
236. Struwe, Fr. E.: Münch. Med. Wschr. 125 (Suppl. 1), 112 (1983)
237. Bahrenfuß, C., Talaulicar, M., Willms, B.: Akt. Endokrin. Stoffw. *5,* 79 (1984)
238. Scarpello, J. H. B.: Diabetes Care *5* (Suppl. 2), 180 (1982)
239. West, T. E. T., Owens, D., Sönksen, P. H.: Clin. Endocrinol. *416,* 573 (1975)
240. Schernthaner, G., Borkenstein, M., Fink, M., Mayr, W. R., Menzel, J., Diabetes Care *6* (Suppl. 1), 43 (1983)
241. Petersen, K. G., Tecklenborg, R., Steinhilber, S. Schlüter, K. J., Burmeister, P., Kasemir, H., Kerp, L.: Immunogenität und Antigenität von biosynthetischem und semisynthetischem Humaninsulin, in: Neue Insuline, (ed.) Petersen et al., p. 159, Freiburg, Freiburger graph. Betriebe (1982)
242. Fireman, P., Fineberg, S. E., Galloway, J. A.: Diabetes Care *5* (Suppl. 2), 119 (1982)
243. Velcovsky, H. G., Federlin, K. F.: Diabetes Care *5* (Suppl. 2), 126 (1982)
244. Fineberg, S. E., Galloway, J. A., Fineberg, N. S., Rathbuon, M. J., Hufferd, S.: Diabetologia *25,* 465 (1983)
245. Heding, L. G., Marshall, M. O., Persson, B., Dahlquist, G., Thalme, B., Lindgren, F., Åkerblom, H. K., Rilva, A., Knip, M., Ludvigsson, J., Stenhammar, L., Strömberg, L., Søvik, O., Baevre, H., Wefring, K., Vidnes, J., Kjaergård, J. J., Bro, P., Kaad, P. H.: Diabetologia *27,* 96 (1984)
246. Fankhauser, S., Herz, G., Zuppinger, K.: Erfahrungen mit Humaninsulin bei diabetischen Kindern, in: Neue Insuline, (ed.) Petersen et al., p. 201, Freiburg, Freiburger graph. Betriebe (1982)
247. Federlin, K., Laube, H., Mäser, E., Velcovsky, H. G.: Münch. Med. Wschr. *125* (Suppl. 1), 101 (1983)
248. Fineberg, S. E., Galloway, J. A., Fineberg, N. S., Rathbuon, M. J.: Diabetes Care *5* (Suppl. 2), 107 (1982)
249. Spijker, A. J., Poortmann, J., Thyssen, J. H. H., Erkelens, D. W.: Diabetes Care *5* (Suppl. 2), 171 (1982)
250. Velcovsky, H. G., Laube, H., Federlin, K.: In Vivo- und in Vitro-Untersuchungen zur Antigenität von verschiedenen Humaninsulinen, in: Neue Insuline, (ed.) Petersen et al., p. 127, Freiburg, Freiburger graph. Betriebe (1982)
250a. Baker, R. S., Schmidtke, J. R., Ross, J. W., Smith, W. C.: Lancet, 1139 (1981)
251. Heding, L. G., Falholt, K., Kristensen, J. S.: IgE Specific for Insulin – Formation and Clincal Significance, in: Immunology in Diabetes (ed.) Andreani, p. 235, London, Edinburgh (1984)
252. Schernthaner, G., Ludwig, H., Jarisch, R., Bruneder, H.: Diabetes Care *4,* 196 (1981)
253. Beischer, W., Maier, V., Jonatha, E. M., Hartweg, D., Kratzsch, G.: Diabetes Care *4* (Suppl. 1), 202 (1981)
254. Kumar, D.: Diabetes Care *4,* 104 (1981)
255. Patterson, R., Mellies, C. J., Roberts, M.: J. Immunol. *110,* 1135 (1973)
256. Falholt, K., Hoskam, J. A. M., Karamanos, B. G., Süsstrunk, H., Viswanathan, M., Heding, L. G.: Diabetes Care *6* (Suppl. 1), 61 (1983)
257. Gries, F. A., Grüneklee, D.: Binding of Biosynthetic Human Insulin (Lilly) to Antibodies Previously Stimulated by Conventional Insulin Therapy, in: Neue Insuline, (ed.) Petersen et

al., p. 135, Freiburg, Freiburger graph. Betriebe (1982)
258. Schernthaner, G.: Diabetes Care *5* (Suppl. 2), 114 (1982)
259. Delvigne, C., Abs, R., de Leeuw, I.: Insulin Allergy Non-responsive to Human Insulin, in: Neue Insuline, (ed.) Petersen et al., p. 149, Freiburg, Freiburger graph. Betriebe (1982)
260. Süsstrunk, H., Morell, B., Froesch, E. R.: Humaninsulin bei Patienten mit Insulin-Allergie, in: Neue Insuline, (ed.) Petersen et al., p. 152, Freiburg, Freiburger graph. Betriebe (1982)
261. Etti, H., Haupt, E.: Akt. Endokr. Stoffw. *4,* 166 (1983)
262. Blandford, R. L., Sewell, R. L., Sharp, P., Hearnshaw, J. R.: Lancet, 1468 (1982)
263. Diem, P., Teuscher, A.: Schweiz. Med. Wschr. *109,* 1814 (1979)
264. Wiles, P. G.: Brit. Med. J. *287,* 531 (1983)
265. Carveth-Johnson, A. D., Mylvaganam, K., Child, D. F.: Lancet, 1287 (1982)
266. Hasche, H.: Allergische Hautreaktionen im Intrakutantest auf sämtliche Humaninsuline bei bestehender Insulinallergie, in: Neue Insuline, (ed.) Petersen et al., p. 158, Freiburg, Freiburger graph. Betriebe (1982)
266a. Bachmann, W., Hasche, H., Mehnert, H.: Diabetes Care *5* (Suppl. 2), 165 (1982)
266b. Holdaway, I. M., Wilson, J. D.: Brit. Med. J. *289,* 1565 (1984)
267. Garcia-Ortega, P., Knobel, H., Mirada, A.: Brit. Med. J. *288,* 1271 (1984)
268. Altman, J. J., Pehnet, M., Slama, G., Tchobroutsky, C.: Lancet, 524 (1983)
269. de Leeuw, I., Delvigne, C., Bekaert, J.: Diabetes Care *5* (Suppl. 2), 168 (1982)
270. Maneschi, F., Fineberg, S. E., Kohner, E. M.: Diabetes Care *5* (Suppl. 2), 175 (1982)
271. Berger, W., Schlüter, K. J.: Biologische Aktivität von Human- und Schweineinsulin bei Rezeptorantikörper-bedingter Insulinresistenz, in: Neue Insuline, (ed.) Petersen et al., p. 227, Freiburg, Freiburger graph. Betriebe (1982)
272. NN, Diabetologie Informationen, Heft 3, p. 12 (1983)
273. Watkins: Brit. Med. J. *287,* 1571 (1983)
274. Sonnenberg, G. E., Berger, M.: Dtsch. Med. Wschr. *108,* 927 (1983)
275. NN, Arzneitelegramm *3,* 19 (1983)
276. Sonnenberg, G. E., Berger, M.: Editorial, Diabetologia *25,* 457 (1983)
277. Karam, J. H.: Diabetes Care *6* (Suppl. 1), 1 (1983)
278. Pfeiffer, E. F.: Münch. Med. Wschr. *125* (Suppl. 1), 121 (1983)
279. Skyler, J. S., Pfeiffer, E. F., Raptis, S., Viberti, G. G.: Diabetes Care *4,* 140 (1981)

Author Index Volumes 1 and 2

The volume numbers are printed in italics

MIX
Papier aus verantwortungsvollen Quellen
Paper from responsible sources
FSC® C105338

If you have any concerns about our products,
you can contact us on
ProductSafety@springernature.com

In case Publisher is established outside the EU,
the EU authorized representative is:
Springer Nature Customer Service Center GmbH
Europaplatz 3, 69115 Heidelberg, Germany

Printed by Libri Plureos GmbH
in Hamburg, Germany